Numerical Analysis II
Lecture Slide Notes

Ralph E. Morganstern
Santa Clara University

Copyright © 2014
By
Ralph E. Morganstern
7504 Phinney Place
San Jose, CA 95139
rmorganstern@yahoo.com

All rights reserved. This work may not be translated or copied in whole or in part without the written permission of the author, except for brief excerpts in connection with reviews or scholarly activity.

ISBN-13: 978-1503071155
ISBN-10: 1503071154

Table of Contents

Preface .... *viii*

1 Initial Value Problem (IVP) *1-1*

1.1 Differential Equations Topics 1-2

1.2 Initial Value Problems - Overview 1-3

1.3 Analytic Solns: Existence & Uniqueness 1-4
1.3.1 Slopefield Plots-1: $f(t,y(t))=-ty^2$ 1-5
1.3.2 MatLab® Quiver Plot: $f(t, y(t)) = -ty^2$ 1-6

1.4 Analytic Solutions: Stability 1-7
1.4.1 Slopefield Plots-2: $f(t,y(t))=-y(1-y)$ 1-8
1.4.2 Slopefield Plots-3: $f(t,y(t))=-\sin(t)/y$ 1-9

1.5 Numerical Solutions: Existence & Uniqueness 1-10
1.5.1 Numerical Solutions: Stability 1-11
1.5.2 Lipschitz Condition for Unique Solution 1-12
1.5.3 Existence, Uniqueness, Stability 1-13

2 Self-Starting Single Step Methods *2-1*

2.1 Euler's Method - Formulation 2-2
2.1.1 Euler's Method - Trajectory 2-3
2.1.2 Euler's Method – 3^d Representation & Slope Surface 2-4
2.1.3 Euler's Method - Perturbations 2-5

2.2 Euler's Method - Error Analysis 2-6
2.2.1 Euler & Higher Order Cumulative Error 2-7
2.2.2 Euler Convergence Theorem 2-8
2.2.3 Euler's Method: Round Off *vs.* Truncation Error 2-9

2.3 Need for Higher Order Methods 2-10
2.3.1 Method-Sampled Slopes 2-11
2.3.2 Sampled Slope Formalism 2-12
2.3.3 Higher Order Taylor Method 2-13
2.3.4 Example: Taylor's Method n =2,3 2-14

2.4 2nd Order Runge-Kutta Derivation 2-15
2.4.1 RK2 Family of Solutions 2-16
2.4.2 RK2 Example 2-17

2.5 RK4 Derivation 2-18
2.5.1 RK4 Derivation – Matched Expansions 2-19
2.5.2 Runge-Kutta Summary & Simpson 2-20

Table of Contents

    2.5.3 RK4 Example – 2 Steps 2-21

2.6 Computational Effort Trade-off: Order vs. Stepsize 2-22
    2.6.1 Computational Effort Trade-off: Example 2-23
    2.6.2 Detailed Comparison of Euler and RK4 2-24

2.7 Error Control – Variable Stepsize 2-25
    2.7.1 RK4 & RKF45 Comparison 2-26
    2.7.2 Runge-Kutta-Fehlberg Method 2-27
    2.7.3 Example of RKF45 with Stepsize Control 2-28

3 *Multistep Methods* *3-1*

3.1 Introduction to Multistep Methods 3-2

3.2 Multistep Formulation – Newton Polynomials 3-3
    3.2.1 Multistep Coefficient Integrals I_k 3-4
    3.2.2 Implicit vs. Explicit Methods 3-5
    3.2.3 Predictor-Corrector Methods – 4th Order 3-6
    3.2.4 Predictor-Corrector Error Control 3-7
    3.2.5 Example of Predictor-Corrector Method with Stepsize Control. 3-8

3.3 Gragg's Extrapolation Technique (Romberg) 3-9
    3.3.1 Gragg's Extrapolation Example 3-10
    3.3.2 Gragg Extrapolation Example Detailed Output Tables 3-11

4 *Systems and Higher Order IVPs* *4-1*

4.1 Coupled Systems of 1st Order Differential Equations 4-2
    4.1.1 Kirchoff System for Currents - Example 4-3

4.2 Higher Order Differential Equations 4-4
    4.2.1 Setup for a 2nd Order ODE 4-5
    4.2.2 2nd Order ODE Example Using RK4 4-6

5 *Special Numerical Considerations for IVP Techniques* *5-1*

5.1 Consistency, Convergence, Stability 5-2
    5.1.1 Examples of Consistency Test 5-3
    5.1.2 Consistency, Convergence, Stability Theorem: Single Step Case 5-4
    5.1.3 Analytic Solutions: Stability 5-5

5.2 Multistep Methods & Parasitic Solutions - 1 5-6
    5.2.1 Multistep Methods & Parasitic Solutions-2 5-7
    5.2.2 Characteristic Equation & Stability 5-8
    5.2.3 Root Structure, Parasitic Solutions, Stability 5-9
    5.2.4 Consistency, Convergence, Stability Theorem: Multi-Step Case 5-10

Table of Contents

    5.2.5 Stability Analysis Examples 5-11

5.3 Stiff Equations 5-12
    5.3.1 Regions of Absolute Stability-General Discussion 5-13
    5.3.2 Stiff Equations - Example 5-14
    5.3.3 Problem with Euler for Stiff IVP 5-15

5.4 Summary of IVP Solutions 5-16
    5.4.1 Implicit Runge-Kutta Family 5-17

6 *Direct Solution of Linear Systems* 6-1

6.1 Preliminaries on Linear Systems 6-2
    6.1.1 Deterministic Solutions 6-3
    6.1.2 Iterative Solutions 6-4
    6.1.3 State Determination from Measurements 6-5
    6.1.4 Round Off Problems Again! 6-6

6.2 Deterministic Solutions 6-7
    6.2.1 Forward Substitution Example 6-8
    6.2.2 Backward Substitution Example 6-9
    6.2.3 Gaussian Elimination Rationale 6-10
    6.2.4 Gaussian Elimination Algorithm 6-11
    6.2.5 Counting Operations: M,A,D 6-12

6.3 Gaussian Elimination Example 6-13
    6.3.1 Revisit Gaussian Elimination 6-14
    6.3.2 Efficiency of $L_1\backslash U$ Factorization 6-15
    6.3.3 Counting: Forward/Backward Elimination 6-16
    6.3.4 Efficient Methods for Special Types of Matrices 6-17

6.4 The Need for Matrix Pivoting 6-18
    6.4.1 Why Scaled Pivoting? 6-19
    6.4.2 Pivoting Overview 6-20
    6.4.3 Partial Pivoting 6-21
    6.4.4 Scaled Pivoting 6-22

6.5 Direct Matrix Inversion 6-23
    6.5.1 Gauss Matrix Inversion 6-24
    6.5.2 Gauss-Jordan Matrix Inversion 6-25

6.6 3^d Determinants 6-26
    6.6.1 N-dimensional Determinants 6-27
    6.6.2 Determinants - Some Properties 6-28

6.7 Other Matrix Factorizations 6-29
6.7.1 DooLittle & CroUt Factorizations 6-30
6.7.2 CholeSki Factorization 6-31
6.7.3 LDU Factorization 6-32
6.7.4 Diagonal Dominant Matrix 6-33
6.7.5 Positive Definite Matrix 6-34

7 *Iterative Solution of Linear Systems* 7-1

7.1 Vector Norms 7-2
7.1.1 Norm Inequalities 7-3

7.2 Matrix Norms 7-4
7.2.1 Convergence of a Vector Sequence 7-5

7.3 Eigenvalues & Eigenvectors 7-6
7.3.1 Eigenvalues & Eigenvectors 7-7
7.3.2 Effect of Matrix A on Arbitrary Vector x 7-8
7.3.3 Computing Eigenvalues/Eigenvectors 7-9
7.3.4 Eigenvector-Eigenvalue Discussion 7-10
7.3.5 Spectral Radius of a Matrix 7-11
7.3.6 Convergent Matrix 7-12

7.4 Motivation for Iterative Methods 7-13
7.4.1 Formulation of Iterative Methods 7-14
7.4.2 Jacobi – Simultaneous Displacement 7-15
7.4.3 Gauss-Seidel–Successive Displacement 7-16
7.4.4 Gauss-Seidel Divergence 7-17
7.4.5 Successive Orthogonal Projections 7-18
7.4.6 Diagonal Dominance & Convergence 7-19
7.4.7 Gauss-Seidel Convergence Theorems 7-20

7.5 Additive Matrix Decomposition 7-21
7.5.1 Jacobi, Gauss-Seidel, Orthogonal Projection 7-22
7.5.2 Convergence and Spectral Radius $\rho(T)$ 7-23

7.6 Accelerating Convergence 7-24
7.6.1 SOR – Math Details 7-25
7.6.2 Example SOR Problem 7-26
7.6.3 Explicit SOR Computation 7-27
7.6.4 Choice of Relaxation Constant ω 7-28
7.6.5 Number of Iterations *vs.* SOR Parameter ω 7-29
7.6.6 Comparison of Iterative Methods: Jacobi, Gauss-Seidel, SOR 7-30
7.6.7 Advantages of Iterative Algorithms 7-31

Table of Contents

7.7 Matrix Condition Number – Gaussian Elimination Issues ... 7-32
    7.7.1 Condition Number Definition ... 7-33
    7.7.2 Estimates of Condition Number ... 7-34

7.8 Iterative Refinement for Gaussian Elimination ... 7-35
    7.8.1 Iterative Refinement Sample Calculation ... 7-36

8 MatLab® Scripts ... 8-1

8.1 Euler Method ... 8-2

8.2 RK4 (Runge-Kutta) Method ... 8-4

8.3 RKF45 (Runge-Kutta-Fehlberg) Method ... 8-6

8.4 Adams AB4-AM3 Predictor-Corrector Method ... 8-8

8.5 Gragg Extrapolation Method ... 8-10

8.6 RK4 for Higher Order or System of ODEs ... 8-12

8.7 Jacobi and Gauss-Seidel Methods for Linear Systems ... 8-14

8.8 SOR Relaxation Methods for Linear Systems ... 8-15

8.9 Comparison of Iterative Methods for Linear Systems ... 8-16

9 References ... 9-1

10 Index ... 10-1

Preface

These Lecture Slide Notes have been used over the past several years for a two-quarter graduate level sequence in numerical analysis. Part 1 covers introductory material on the Nature of Numerical Analysis, Root Finding Techniques, Polynomial Interpolation, Derivatives, and Integrals. Part 2 covers Ordinary Differential Equations and Numerical solutions to Linear Systems of Equations.

This "Lecture Slide Notes" format is convenient for self-study because it covers the subject matter in a concise and easily accessible manner by employing multiple visualization techniques in slide format together with focused explanatory notes. Each slide stands alone as a "one-page synopsis" that encapsulates a complete concept, algorithm, or theorem using a combination of equations, graphs, diagrams, plots, illustrative tableaus and comparison tables. The explanatory notes are placed directly below each slide in order to reinforce and/or give additional insight into the particular numerical technique or concept illustrated in the slide.

Part 2 starts with a discussion of the fundamental initial value problem (IVP) for a first order ordinary differential equation (ODE) $y' = f(t, y(t))$ and includes both single and multi-step methods with stepsize control. The single step methods discussed are the 1st order Euler Method, n^{th} order Taylor methods, general Sampled Slope methods, Runge-Kutta-Fehlberg methods and finally an Extrapolation method. Multi-step methods discussed are the *explicit* Adams-Bashforth 4-step and *implicit* Adams-Moulton 3-step methods as well as their Predictor-Corrector combination. Algorithm consistency, convergence, and stability and the extension to systems and higher order ODEs and to stiff DEs is also discussed in some detail.

The second half of this book covers direct (deterministic) and iterative solutions to n x n linear systems of equations. Both direct matrix inversion and linear system solution of the fundamental matrix equation **Ax=b** are explored. Topics covered are Gaussian elimination, counting operations, pivoting strategies, determinants and direct matrix inversion. The role of matrix decompositions and efficient methods for special matrices is also discussed. Iterative methods covered are Jacobi, Gauss-Seidel, Orthogonal Projections, and SOR relaxation; a simple "additive" decomposition of the matrix **A** is used to cast them into a common framework for comparison. Finally a brief discussion of matrix condition number K(**A**) and the role of iterative refinement for Gaussian elimination is given together with some explicit examples.

A Table of Contents serves to organize the slides in terms of the main numerical analysis topics covered and gives a complete list of slide titles and their page numbers. An index is also provided to link related aspects of topics and cross-reference key concepts, specific applications, and the various visualization aids. Although no problem sets have been included in these notes, a good number of examples are worked out in detail. Moreover, Section 8 contains a selection of illustrative MatLab® scripts that are keyed to these examples and can easily be edited by the student to generate solution tables and plots for similar problems. Finally, references to a number of standard text books are given, but there has been no attempt to make an exhaustive bibliography.

1 Initial Value Problem (IVP)

Initial Value Problem (IVP)

1.1 Differential Equations Topics

Differential Equations Topics

1. Initial Value Problems (1st Order DE s)
2. Existence, Uniqueness, Stability
3. Euler's Technique & Error Analysis
4. Runge-Kutta Self-Starting Techniques
5. Stepsize & Error Control
6. Multistep Methods (Non-self Starting)
7. Predictor-Corrector Methods
8. Extrapolation Techniques
9. Higher Order DE & Systems
10. Algorithm Stability & Parasitic Solutions
11. Stiff Differential Equations

We shall be considering, almost exclusively, a 1st Order Ordinary Differential Equation that equates the time derivative of y(t) to an arbitrary "slope function" f(t,y) which depends upon both the time t and y itself. This differential equation together with a single start value $y(t=a)=\alpha$ is the fundamental Initial Value Problem (IVP) that we will spend the 1st half of the course on. The methods used to solve this simple IVP problem are easily extended to higher order DEs as well as systems of coupled DEs.

Here we enumerate a rich set of topics related to the numerical solution of this fundamental IVP. As with any numerical technique the existence, uniqueness, and stability of solutions, as well as the solution precision must be related to the specific computational algorithm and to the ever-present trade-off between truncation and round-off errors.

The Euler method provides a 1st order solution that is easily analyzed and is the simplest self-starting method that proceeds from the given initial condition to give a step-by-step integration of the differential equation. Higher order Runge-Kutta methods improve the computational efficiency by using fewer steps to reach a given coordinate thereby reducing the cummulative round-off error. Also, by appropriate stepsize changes, the numerical solution maintains its accuracy over large coordinate runs even as the solution varies rapidly over time.

Multistep methods are not self starting because initial condition must be augmented by several other step values in order to proceed. Predictor-Corrector and Extrapolation methods are discussed in this context. Multistep algorithms produce parasitic solutions and require special care to insure solution consistency, convergence, and stability; they lead to the concept of stiff differential equations and the associated implicit solution techniques.

Initial Value Problem (IVP)

1.2 Initial Value Problems - Overview

Initial Value Problems - Overview

- **1st Order DE – Fundamental IVP**

$$y' \equiv \frac{dy}{dt} = f(t, y(t)) \quad t \in [a,b] \quad IC: y(t=a) = \alpha$$

- **System 1st Order Eqns:** *Vectorize*

$$\frac{dy_1}{dt} = f_1(t, y_1(t), y_2(t))$$
$$\frac{dy_2}{dt} = f_2(t, y_1(t), y_2(t))$$
$$t \in [a,b] \quad IC: \begin{matrix} y_1(t=a) = \alpha_1 \\ y_2(t=a) = \alpha_2 \end{matrix}$$

$$\frac{d\bar{y}}{dt} = \bar{f}(t, \bar{y}(t)) \quad t \in [a,b] \quad IC: \bar{y}(t=a) = \bar{\alpha}$$

$$\bar{y} \equiv \begin{bmatrix} y_1(t) \\ y_2(t) \end{bmatrix} \;;\; \bar{\alpha} = \begin{bmatrix} \alpha_1 \\ \alpha_2 \end{bmatrix} \;;\; \bar{f} \equiv \begin{bmatrix} f_1(t, \bar{y}(t)) \\ f_2(t, \bar{y}(t)) \end{bmatrix}$$

- **2nd Order DE** *Vectorize*

$$\frac{d^2 y}{dt^2} = y'' = f(t, y(t), y'(t)) \quad t \in [a,b]$$
$$IC: y(t=a) = \alpha \;;\; y'(t=a) = \beta$$

$$\frac{d\bar{y}}{dt} = \bar{f}(t, \bar{y}(t)) \quad t \in [a,b] \quad IC: \bar{y}(t=a) = \bar{\alpha}$$

$$\bar{y} \equiv \begin{bmatrix} y_1(t) \\ y_2(t) \end{bmatrix} = \begin{bmatrix} y(t) \\ y'(t) \end{bmatrix} \;;\; \bar{\alpha} = \begin{bmatrix} \alpha \\ \beta \end{bmatrix} \;;\; \bar{f} \equiv \begin{bmatrix} y_2(t) \\ f(t, \bar{y}(t)) \end{bmatrix}$$

$$\begin{bmatrix} y_1 = y \;;\; y_2 = y' \\ y_2' = y'' = f(t, y_1(t), y_2(t)) \;;\; y_1' = y_2 \\ IC: y_1(t=a) = \alpha \;;\; y_2(t=a) = \beta \end{bmatrix}$$

The fundamental IVP is specified by three elements, namely (i) the ordinary differential equation (DE): $y'(t) = f(t,y(t))$, (ii) the time domain $t \in [a,b]$, and (iii) the initial condition $y(t=a) = \alpha$. Note that the DE is geometrically interpreted as a specification of the "slope function" $f(t,y(t))$ for the time derivative $y'(t)=dy/dt$ for all points in the 2^d domain $D = \{t \in [a,b], y \in (-\infty, +\infty)\}$; $f(t,y(t))$ is a surface above the t-y plane which determines the slope for every value of t and y in D.

The importance of studying this IVP is that it may be extended to both n^{th} order DEs and systems of n-coupled 1st order DEs. Specifically, for a system of two 1st order DEs, we may write the IVP for the system equations in terms of the 2-vectors **y**, **f** and **α** (bolded) as displayed in the upper Vectorize Box on the slide. The detailed structure consists of

(i) the **vectorized variables**:
$$\mathbf{y}(t) = [y_1(t), y_2(t)]^T \;;\; \mathbf{y'}(t) = [y_1'(t), y_2'(t)]^T$$
$$\mathbf{f}(t,\mathbf{y}(t)) = [f_1(t, y_1(t), y_2(t)), f_2(t, y_1(t), y_2(t))]^T \text{ where superscript T is the transpose, and}$$

(ii) the **two initial conditions**:
$$\mathbf{y}(t=a) = [y_1(t), y_2(t)]^T = \boldsymbol{\alpha} = [\alpha_1, \alpha_2]$$

In this compact notation we have a vector equation and a vector initial condition that have the same form as in our original **scalar IVP**; each component satisfies the same "form" DE and IC.

In a similar manner, the components in the 2nd order DE are defined by y and its first derivative y' as $y_1=y$ and $y_2=y'$. This case also yields the identical vectorized equation as given by the lower Vectorize Box in the slide. Thus, methods developed for this IVP are applicable to a wide range of problems.

1.3 Analytic Solns: Existence & Uniqueness

Analytic Solns: Existence & Uniqueness

- **Singularity in analytic solution**
  - Unique solution not guaranteed across a singularity in either $y(t)$ or $f(t, y(t))$
  - Require solution, $y(t)$ & slope fcn $f(t, y(t))$ to exist and be continuous for *all points in the interval* $t \in [a,b]$
- **Example:**

$$\frac{dy}{dt} = -ty^2 \quad t \in [0,4]$$

IC1: $y(\mathbf{0}) = 1 \Rightarrow y = \dfrac{2}{t^2 + 2}$

IC2: $y(\mathbf{2}) = 1 \Rightarrow y = \dfrac{2}{t^2 - 2}$

$$-\int \frac{dy}{y^2} = \int t\, dt \Rightarrow \frac{1}{y} = \frac{t^2}{2} + C$$

$$y = \frac{2}{t^2 + 2C}$$

- **Solution not unique for IC2: No way of matching across the singularity**

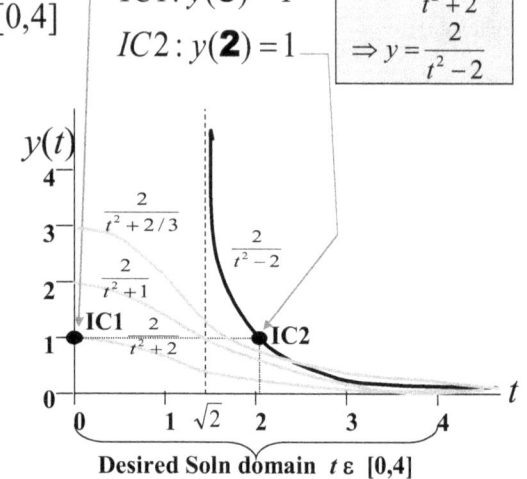

Desired Soln domain $t \varepsilon\ [0,4]$

Even before we consider issues related to the existence and uniqueness of numerical solutions for a given IVP, we must first insure that the analytic solution itself exists and is unique.

It is important to understand that the IVP consist of three parts, namely, (i) the differential equation DE, and (ii) the initial condition IC and (iii) the time domain t∈[a,b] for which the solution is to be valid. It is entirely possible to have a unique analytic solution for one domain and IC, but not for another as the example on this slide shows.

In the example, the DE y'(t) =-ty^2 and time domain t ∈ [0,4] are fixed and we simply change the initial conditions to yield the two very different solutions:

IC1: y(0)=1 yields y = 2/[t^2 + 2] and IC2: y(2)=1 yields y = 2/[t^2 - 2]

As shown in the plot the first solution for IC1 starts at the grey dot (y(0)=1) is one of a family of solutions with ICs y(0)=1, 2, 3, that are all well behaved and unique over the entire domain t∈[0,4]. On the other hand, the second solution for IC2 starts at the black dot (y(2)=1) and is well behaved only over part t∈[(2)^{½},4] of the desired domain t∈[0,4] because of the singularity in the denominator [occurring when t^2 - 2 =0]. Because of this singularity at t = (2)^{½}, there is no unique way to match it to solutions for t < (2)^{½} and hence there is no valid solution to the IVP with IC2.

Thus, knowing that the analytic problem does not have a solution precludes an attempt at finding a numerical solution.

Initial Value Problem (IVP)

1.3.1 Slopefield Plots-1: $f(t,y(t))=-ty^2$

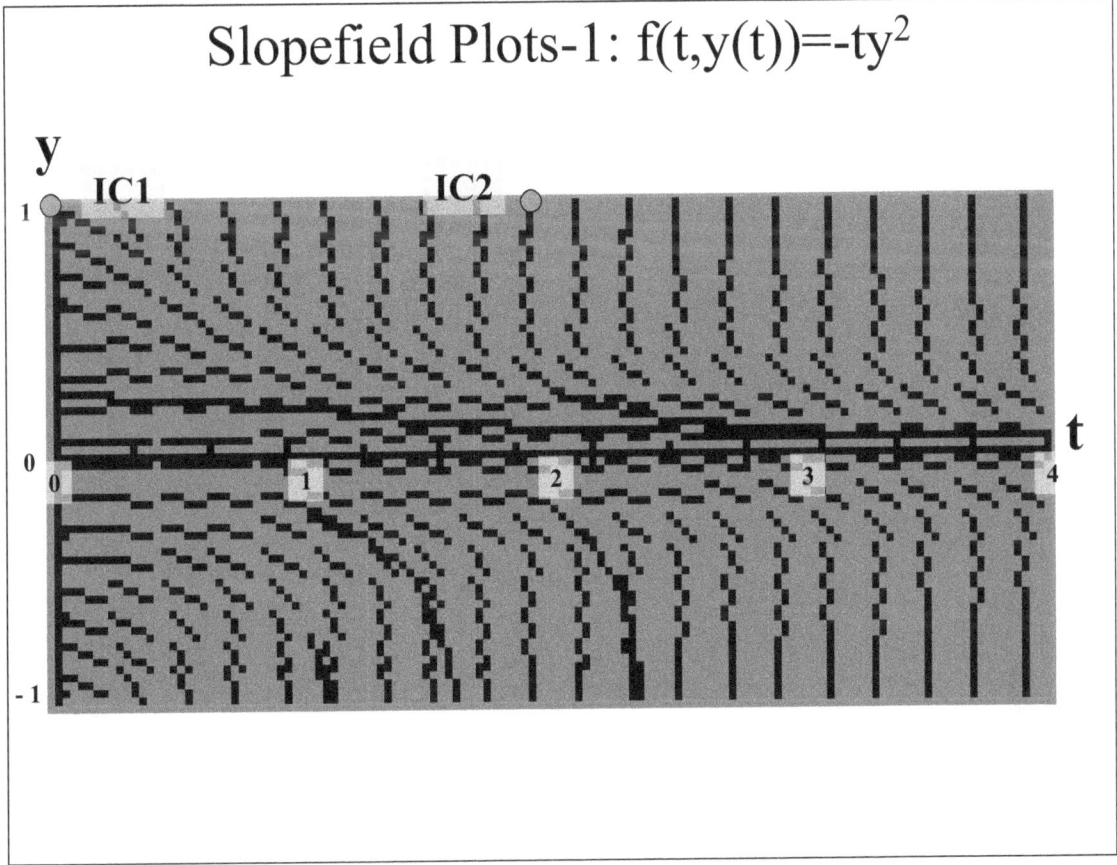

As we have seen, the existence and uniqueness of a solution to an IVP over a given domain $t \in [0,4]$ may depend upon the choice of IC. One method of visualizing the nature of all solutions of an IVP is to plot the value of the slope over a grid of points in the t-y plane. The plot in the slide was performed on an HP 48G series calculator using the native "slope field" plot option for the DE $y' = f(x,y(x)) = -xy^2$ (where x replaces t on the calculator).

The slope at each point (t,y) is represented by the short (jagged) line segment centered on each grid point; the grid of points themselves are not visible. The shape of the solution curve starting from any initial condition can be visualized by viewing the plot from afar and tracing by eye a continuous curve through the slope segments. The HP 48G calculator can overlay the solutions for various initial conditions by using the native "Differential Equation" plot routine and this is done for several different ICs as seen by the "darkened" lines in the figure.

Unfortunately the scales are not very accurate on this screen capture of the HP 48G plot, but are available on the calculator itself. However, we can eyeball solution curves for ICs $y(2)=1$, $y(3)=1$,... that display the singular behavior and go to $+\infty$ at the vertical asymptotes such as $t = (2)^{1/2}$, $(7)^{1/2}$, ...as well as those that have ICs $y(0)=1$, $y(0)=2$, ... that are well-behaved over the whole domain $t \in [0,4]$.

These plots should be performed in MatLab® for more visually pleasing displays and analytics. A very simple MatLab® script and plot is given on the next slide.

Initial Value Problem (IVP)

1.3.2 MatLab® Quiver Plot: $f(t, y(t)) = -ty^2$

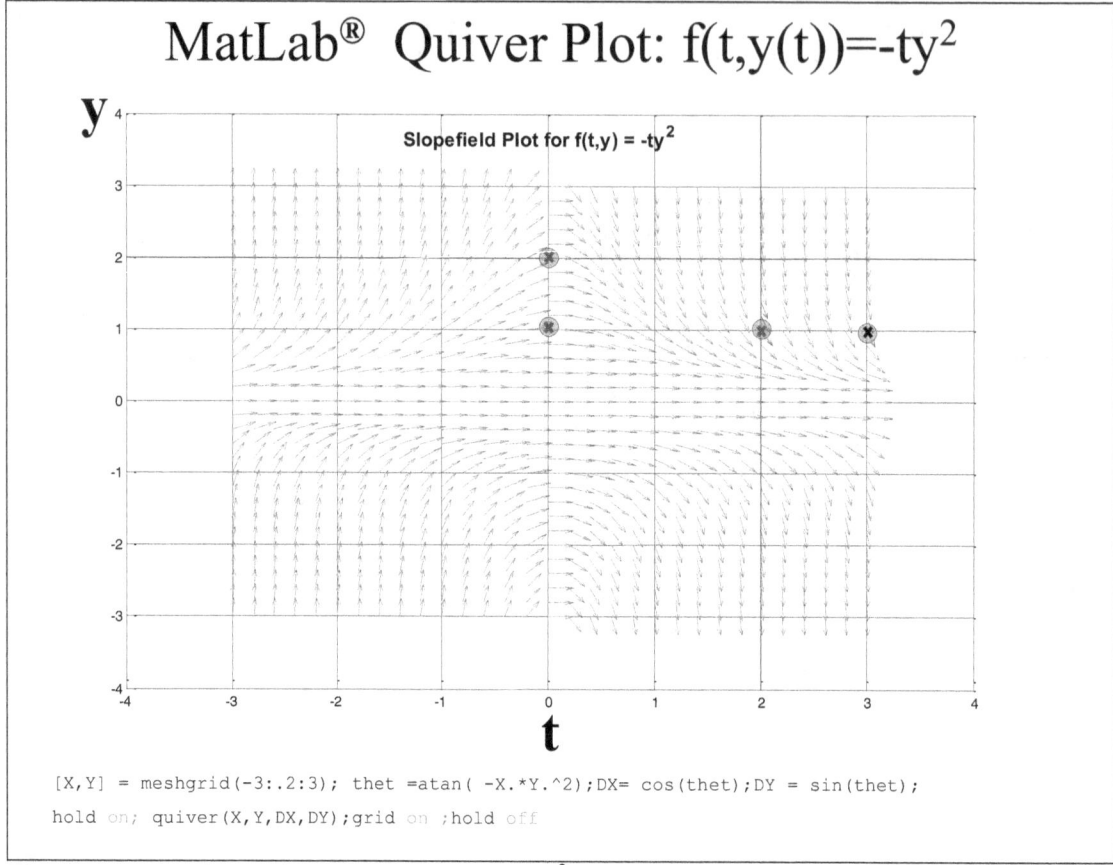

The same DE as in the last slide $y' = f(t,y(t)) = -ty^2$ is plotted for $t \in [-3,3]$ with the simple **MatLab® Script** given below. The slope at each point (t,y) is represented by the short arrow centered on each grid point using the native "quiver" plot option. The shape of solution curves for any given initial condition is easily visualized by viewing the plot from afar and tracing by eye a continuous curve through the slope segments.

We can eyeball solution curves starting at the two points ⊛ shown on the y-axis corresponding to the ICs y(0)=1, y(0)=2 and verify that they are well-behaved over the whole domain $t \in [0,3]$. Similarly we can visualize the solution trajectories for the ICs y(2)=1, y(3)=1 which display a singular behavior and go to $+\infty$ at the vertical asymptotes $t = (2)^{1/2}$, $(7)^{1/2}$ respectively (see Slide#1-4). These latter curves do not cover the desired domain $t \in [0,4]$ and hence do not constitute a valid solution of the IVP.

MatLab® Script

[X,Y] = meshgrid(-3:.2:3);	% sets up square mesh grid
thet =atan(-X.*Y.^2);	% compute angle with x-axis at each point
DX= cos(thet);	% X component of slope for quiver plot arrow
DY = sin(thet);	% Y component of slope quiver plot arrow
hold on;	% Leaves figure open for additional plots
quiver(X,Y,DX,DY);	% **MatLab®** quiver plot places scaled arrows at each grid point
grid on ;	% turns plot grid on
hold off	% closes figure

Initial Value Problem (IVP)

1.4 Analytic Solutions: Stability

Analytic Solutions: Stability

- **Standard of Stability**

$$y' = Ay \;;\; y(0) = \alpha$$

$$y = \alpha e^{At}$$

Stable to small perturbations

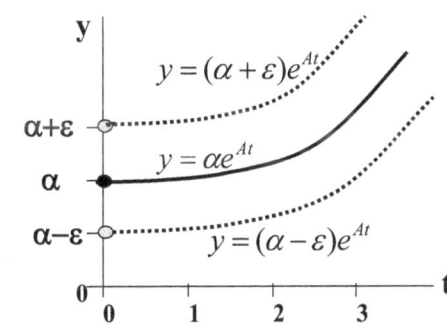

- **Standard of Instability**

$$y' = -y(1-y) \;;\; y(0) = \alpha$$

$$y = \frac{\alpha}{\alpha + (1-\alpha)e^t}$$

Unstable for $\alpha = 1 \pm \varepsilon$
Solution changes radically
For small perturbations
about $\alpha = 1$

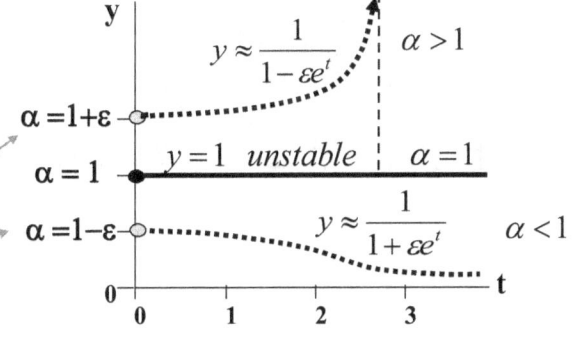

We have seen in the slope plot that besides the issues of existence and uniqueness over a specified domain, the solution may become unstable and approach infinity for some specifications of the initial condition (IC). Even though this may be unavoidable over large changes in IC, we would like to insure that it does not happen when there is a **small perturbation** ε on an initial condition from say $\alpha = 1$ to $1+\varepsilon$ or $1-\varepsilon$. This property is extremely important because in numerical solutions there is always a small perturbation by virtue of the limited precision on conversion from floating point to machine numbers; that is, the analytic solution may be fine at a given $y(0) = 1$, but the small perturbation $y(0) = \text{float}(1) = 1+\varepsilon$ (or $1-\varepsilon$) caused by conversion of this IC to a machine represented number may yield a divergent solution.

The two DEs on this slide are subjected to small perturbations $y(0) = \alpha \pm \varepsilon$ in their ICs to test their stability. Their solutions are given in the top and bottom panels of this slide and designated as the *standard of stability*, and the *standard of instability*, respectively. The top figure illustrates that the perturbed solutions "track" the analytic solution by remaining close to it for the stable exponential case. The bottom figure illustrates that for the instability standard, the perturbed solutions do not always "track" the analytic solution; more specifically, for the specific case $\alpha=1$, the analytic solution is simply a constant $y=1$, but the two perturbed solutions do not track the analytic solution and in fact have completely different behaviors with $y(0) = 1+\varepsilon$ approaching infinity and $y(0) = 1-\varepsilon$ approaching zero asymptotically. Clearly the constant, $y=1$ solution is unstable to small perturbations.

Note that the solution to the unstable DE is obtained by separating variables and making a partial fraction expansion to obtain $-dt = dy/[y(1-y)] = dy/y + dy/(1-y)$ which integrates to give

$$-t + \ln(C) = \ln(y) - \ln(1-y) = \ln(y/(1-y)) \rightarrow y/(1-y) = Ce^{-t} \text{ or } y = Ce^{-t}/[1+Ce^{-t}] = C/[C+e^t]$$

Applying the IC $y(0) = \alpha = C/[1+C]$ yields $C = \alpha/(1-\alpha)$ and the solution $y = \alpha / [\alpha + (1-\alpha) e^t]$

Initial Value Problem (IVP)

1.4.1 Slopefield Plots-2: f(t,y(t))=-y(1-y)

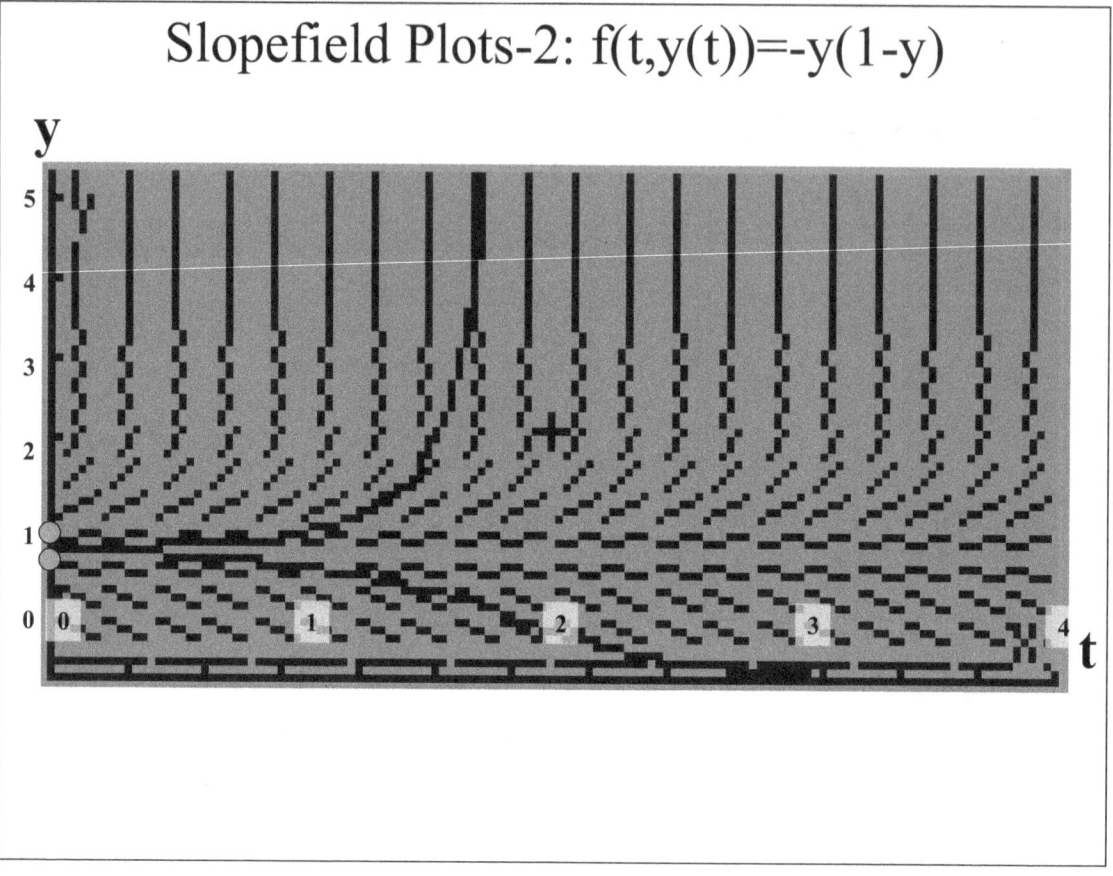

Here is an HP 48GX screen shot of a slope field plot for the *Standard of Instability* DE. We have overlaid the unperturbed and perturbed solutions for the case $y(0) = \alpha = 1$, namely
(i) $\alpha = 1$, yields the unperturbed constant solution $y=1$,
(ii) $\alpha = 1- \varepsilon$, yields the perturbed divergent solution $y = 1/[1- \varepsilon\, e^t]$ with asymptote at $t = -\ln(\varepsilon)$, and
(iii) $\alpha = 1+\varepsilon$, yields the perturbed solution $y = 1/[1+ \varepsilon\, e^t]$ converging to zero.
The slope field plot makes it clear that $\alpha = 1$ is a unique initial condition inasmuch as α-values above 1 all solutions diverge to infinity with different asymptotes (vertical lines) and those α-values below 1 all converge to zero. Thus the plot allows us to easily visualize the instability of the solution for small perturbations about $\alpha = 1$ as they clearly do not track the unperturbed constant solution $y=1$.

Initial Value Problem (IVP)

1.4.2 Slopefield Plots-3 : f(t,y(t))=-sin(t)/y

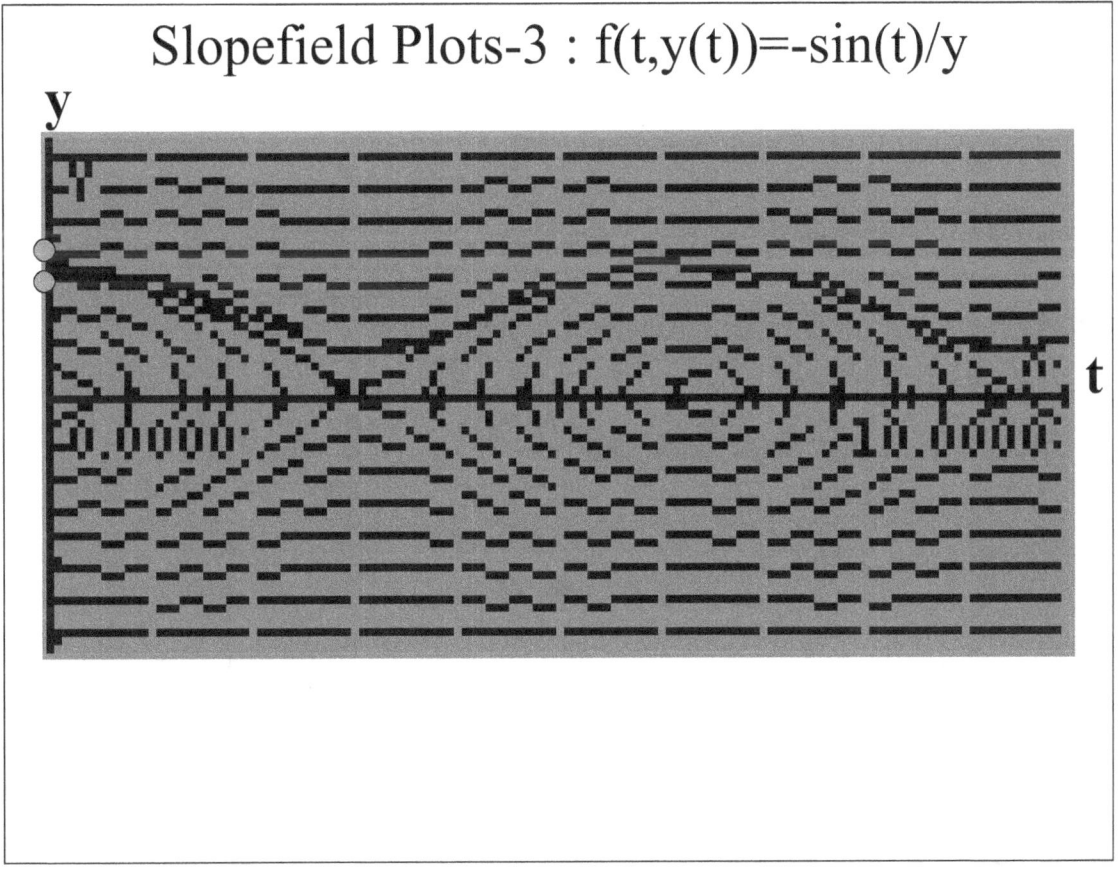

Here is another HP 48GX screen shot of a slope field plot for the function $f(t, y(t)) = -\sin(t)/y$ over the domain D: $t \in [0,10]$ and $y \in [-1.5, 1.5]$. The pattern of slopes combines the sinusoidal variation along the t axis with the $1/y$ behavior along the y-axis and illustrates two basic types of trajectory with start at the two points shown at $y(0) = \alpha$, namely, (i) the upper point for which the solution trajectory is a sinusoidal type curve having only positive y-values covering the full time interval [0, 10] and (ii) the lower point whose sinusoidal trajectory stops at the x-axis and cannot proceed further; the "trapped trajectory" appears to cross the t-axis and continue back to t=0 and a negative value of y. The bottom half of the trajectory actually corresponds to an initial value at the mirror point $y(0) = -\alpha$ and ends on the axis at $t = \pi/2$. Note there is a critical trajectory between these two points which is a true sinusoid taking on both positive and negative values as it covers the full range [0, 10]. It is clear that this critical trajectory separates the IVP solutions into trapped with limited range and full trajectories with the unlimited range of a sinusoid.

It would be nice to know under what conditions the solution will be unique and stable prior to attempting a numerical solution to the IVP. These plots are one method, but they require decisions based on visual data; the theorem giving the mathematical conditions for a unique stable solution is discussed on the next few slides.

1.5 Numerical Solutions: Existence & Uniqueness

Numerical Solutions: Existence & Uniqueness

- 1st Order DE – Fundamental IVP

$$y' \equiv \frac{dy}{dt} = f(t, y(t)) \quad t \in [a,b] \qquad IC: \; y(t=a) = \alpha$$

- Curve $y(t)$ is tangent to slopes $= f(t, y(t))$
- Starting with IC

  at $t_0 = a$ slope is: $f(a, \alpha)$

  When is there a unique soln curve?

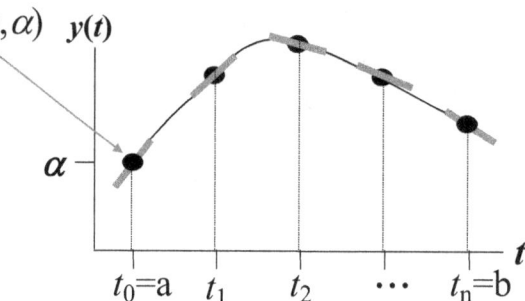

To solve the IVP, we start at the initial time $t_0 = a$ with a y-value $y(t_0) = \alpha$ which is shown as the 1st black dot in the figure. We can compute the slope at the start point by evaluating the slope function at $f(a,\alpha)$ and this value is represented by a red slope segment at an angle with the t-axis determined by θ = arctan[$f(a,\alpha)$] ; this slope segment is the tangent to the unknown solution curve. If we actually knew the solution, we could evaluate the slope function at each point along the solution curve and draw small red slope segments at each of these points as shown in the figure.

It would be nice to know precisely under what conditions a unique solution to the IVP exists prior to expending efforts to find a numerical solution.

Note that the *slope field plots* on the previous slides show all the slopes on an evaluation grid and allow one to choose a specific *initial condition* and *eyeball* the solution curve by kissing the slopes as we move through the grid.

1.5.1 Numerical Solutions: Stability

Numerical Solutions: Stability

- Even when analytic soln is stable to IC
  - Numerical method may introduce instabilities
  - Small perturbations in initial conditions & slope $f(t,y)$ can yield a drastic change in character of solution.
- What conditions yield a unique & stable soln?

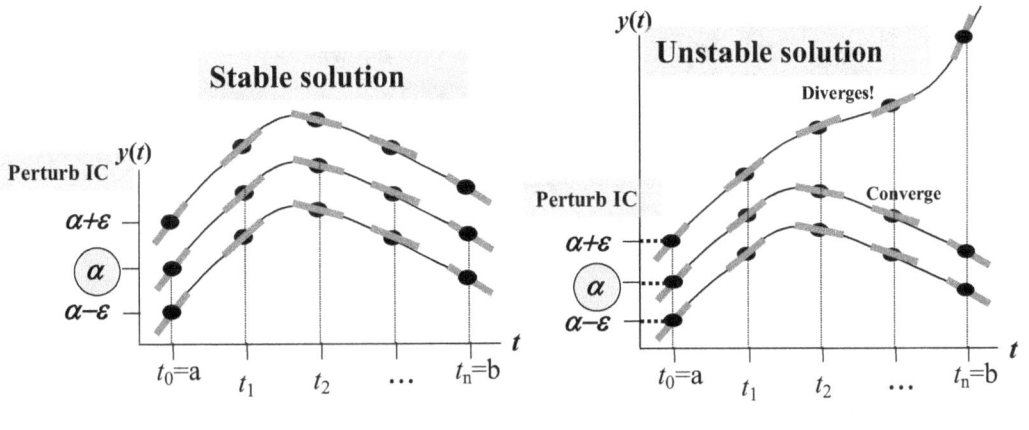

Even if a unique analytic solution exists and is stable for a given IVP, we are not guaranteed that a particular numerical algorithm will have these same characteristics. In fact, numerical algorithms often generate "parasitic solutions" that diverge from the true solution. This can occur when the numerical algorithm involves the solution of a 2nd (or higher) order difference equation whose associated characteristic polynomial has roots with magnitude greater than "1"

Because of round off errors numerical solutions always introduce small perturbations to the initial condition $y(t=a) = \alpha$ and/or the slope function $f(t, y(t))$ which in turn introduce small multipliers to the parasitic terms when the initial conditions are applied. Even with very small coefficients parasitic terms corresponding to roots greater than 1 will eventually dominate the true solution and cause divergence as illustrated in the figure on the right. The left hand figure shows a stable situation in which the slopes on neighboring trajectories track one another quite well, while the right hand figure shows a hypothetical situation for which the lower trajectory is OK, while the upper trajectory yields a divergent and unrelated solution.

Note that this perturbation causes an error in the computation of the slope function either because (i) the estimated y-value does not lie on the true curve or (ii) the computation of the slope itself produces round-off errors.

Initial Value Problem (IVP)

1.5.2 Lipschitz Condition for Unique Solution

Lifschitz Condition for Unique Solution

- **Lifschitz Condition:** $|f(t,y_1) - f(t,y_2)| \leq L \cdot \underbrace{|y_1 - y_2|}_{\text{Diff. 2}^{\text{nd}}\text{ Var}}$ on set D
  - **Example:**
  $$\boxed{f(t,y) = t|y| \quad t \in [1,2] \; ; \; y \in [-3,4]}$$
  $$|f(t,y_1) - f(t,y_2)| = \left| \{t \cdot |y_1| - t \cdot |y_2|\} \right| \leq |t| \cdot |y_1 - y_2| \leq 2|y_1 - y_2|$$

- **Convex Set**: St. line drawn btwn two pts in set remains in set

  $(t_1, y_1), (t_2, y_2) \in D$ iff $(t(\lambda), y(\lambda)) \in D \; ; \; \lambda \in [0,1]$

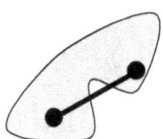
Not Convex!

Parametric form of St. Line btwn Pts
$$t(\lambda) = (1-\lambda)t_1 + \lambda t_2$$
$$y(\lambda) = (1-\lambda)y_1 + \lambda y_2 \Rightarrow \frac{dy}{dt} = \frac{dy/d\lambda}{dt/d\lambda} = \frac{(y_2 - y_1)}{(t_2 - t_1)}$$

Yes Convex!

$$\boxed{D \equiv \{(t,y) : t \in [a,b], y \in (-\infty, +\infty)\}}$$

- **Thm**: Computing Lifschitz constant L

  If $f(t,y)$ defined on convex set D

  and $\left|\dfrac{\partial f(t,y)}{\partial y}\right| \leq L$ for all $(t,y) \in D$

  Then $|f(t,y_1) - f(t,y_2)| \leq L \cdot |y_1 - y_2|$

From the discussion of slope field plots and their ability to visually summarize the nature of the solutions to an IVP, it should not be surprising that f(t,y(t)) plays a key role in determining a unique solution.

First we note that the solution domain D≡{(t,y): t∈[a,b], y ∈ (- ∞ , +∞) } is a Convex set; this concept requires the line between arbitrary two points in the set remain in the set D as illustrated in the inset figure. The Lipschitz condition on f(t,y) puts a bound on the absolute difference of the slope function for all pairs of points (t,y₁) and (t, y₂) in D at a fixed time t as follows
| f(t,y₁) - f(t,y₂) | ≤ L| y₁ - y₂ | for all points in the convex domain D
In the example f(t,y) = -t|y| in the restricted domain t∈[1,2]; y∈[-3,4], the Lipschitz condition is yields
 | f(t, y₁) - f(t, y₂) | =| t|y₁| -t|y₂||= t| |y₁|-|y₂| | ≤max(t) | y₁ − y₂ | = 2 |y₁ − y₂ | ➔ L=2
Note that in the unrestricted domain D the difference approaches ∞, and the Lipschitz condition does not hold.
The theorem at the bottom of the page states the equivalence between | f(t,y₁) - f(t,y₂) | ≤ L| y₁ - y₂ | and the inequality for the derivative |fy(t,y)| ≤ L, which is a simpler way to determine whether the Lipschitz condition holds. Thus taking the partial with respect to y, is equivalent to the Liftschitz condition and for the above example we have directly
 |∂/∂y(-t·|y|)| =max(t) =2 =L.

Initial Value Problem (IVP)

1.5.3 Existence, Uniqueness, Stability

Existence, Uniqueness, Stability

- **Thm: Existence & Uniqueness** (Sufficient Conditions)

  If $f(t,y) \in C$ on convex set D
  and Lipschitz: $|f(t,y_1) - f(t,y_2)| \leq L \cdot |y_1 - y_2|$ for all $(t,y) \in D$
  Then IVP $y'(t) = f(t, y(t))$; $t \in [a,b]$; $y \in (-\infty, +\infty)$
  has unique solution $y(t)$

  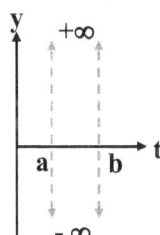

- **Stability to Perturbations** (Well-posed IVP)

  IVP: $y' = f(t,y)$; $y(a) = \alpha$; $t \in [a,b]$
  $IVP_{Perturbed}$: $z' = f(t,z) + \delta(t)$; $z(a) = \alpha + \varepsilon_0$; $t \in [a,b]$
  $|z(t) - y(t)| < k\varepsilon$; $t \in [a,b]$ where $\varepsilon = \max\{\varepsilon_0, \delta\}$ & $k > 0$

- **Thm: Well-posed IVP**

  If IVP satisfies Lifschitz condition and $f(t,y) \varepsilon C$ (continuous on D), **then** it is well-posed

-

The **sufficient conditions** for the existence and uniqueness of a solution to an IVP is that the function be continuous and satisfy a Lipschitz condition on the convex set D as given by the 1st theorem. Note that these conditions are sufficient but not necessary; i.e., it is possible for a unique solution to exist even if these two conditions are not satisfied. The theorem only states that satisfaction of the two conditions guarantees the existence of a unique solution.

A **Well-posed IVP** has a stable solution under small perturbations both in the initial condition and the slope function. The perturbed IVP is defined explicitly in terms of a new solution variable "z" and the stability of the solution requires z(t) to "track" y(t), which is expressed mathematically by requiring the absolute difference |z(t) - y(t)| < k ε   for all t∈[a,b] where ε represents the maximum perturbation, and k is a positive number less than one.

The 2nd Theorem states that the IVP is well-posed provided continuity and Lipschitz condition hold for f(t,y) over the domain D. .

2 Self-Starting Single Step Methods

2.1 Euler's Method -Formulation

Euler's Method -Formulation

- 1st Order DE – Fundamental IVP:

$$y' \equiv \frac{dy}{dt} = f(t, y(t)) \quad t \in [a,b] \quad IC: y(t=a) = \alpha$$

- Mesh: $t_i = a + ih \quad i = 0, 1, \ldots N \quad (N+1 \text{ pts}, N \text{ intvls})$
- Step: $h = (b-a)/N$
- Taylor: $y(t_{i+1}) = y(t_i) + y'(t_i) \cdot h + y''(\xi) \cdot \frac{h^2}{2} \quad \xi \in [t_i, t_{i+1}]$
- Subs. D.E.: $y(t_{i+1}) = y(t_i) + \overbrace{f(t_i, y(t_i))} \cdot h + y''(\xi) \cdot \frac{h^2}{2}$
- Replace & Generate Approx Soln $w_i \cong y(t_i)$
- Associated Difference Eqn:

$$w_{i+1} = w_i + f(t_i, w_i) \cdot h$$

- Initial Condition: $w_0 = y(a) = \alpha \quad [\text{Note}: w_i \neq y(t_i) \text{ for } i \neq 0]$

In order to solve an IVP, we set up N time-steps of size h =(b-a)/N in the interval [a,b]. (Note that this means there are N+1 time points t_i = a + i*h where i=0,1,..., N. The simplest approach (Euler Method) is to make a 1st order Taylor expansion of the solution $y(t_1)$ about the point t = t_0, viz.,

$y(t_1) = y(t_0) + y'(t_0)*h + y''(\xi)* h^2/2$ for $\xi \in [t_0, t_1]$ with $t_1 - t_0 = h$

Using the DE $y'(t_0) = f(t_0, y(t_0))$ in the above equation yields

$y(t_1) = y(t_0) + f(t_0, y(t_0)) h + y''(\xi)* h^2/2$ with IC: $y(t_0) = \alpha$

which is an exact expression for $y(t_1)$ including the Taylor error term. If we drop this unknown error term and re-label $y(t_0) = w_0 = \alpha$ and $y(t_1) = w_1$ this yields a difference equation which approximates the original differential equation

$w_1 = w_0 + f(t_0, w_0) h$ with IC: $w_0 = \alpha$ (which is exactly equal to $y(t_0)$)

Note that in making this transition to a difference equation we have dropped the error term and replaced y_1 with w_1 making it an exact equation again now in terms of the "w"s. The result can be interpreted geometrically as the sum $w_1 = w_0 + \Delta w_0$ where Δw_0 is just the "Euler slope" times the horizontal step "h". The process continues in an obvious manner to obtain $w_2 = w_1 + f(t_1, w_1) h$ and the general result for w_{i+1} is given in the boxed equation. We note that only $w_0 = \alpha = y(t_0) = y_0$ exactly; for all other values of "i" we only have approximate equality $w_1 \approx y_1$, $w_2 \approx y_2$ etc. . The resulting sequence of points $\{w_0, w_1, w_2, \ldots\}$ generated by the difference equation do not coincide with the sampled y-values $\{y_0, y_1, y_2, \ldots\}$ except at the initial point where $w_0 = y_0$. This Euler integration technique is illustrated in the next slide.

Self-Starting Single Step Methods

2.1.1 Euler's Method - Trajectory

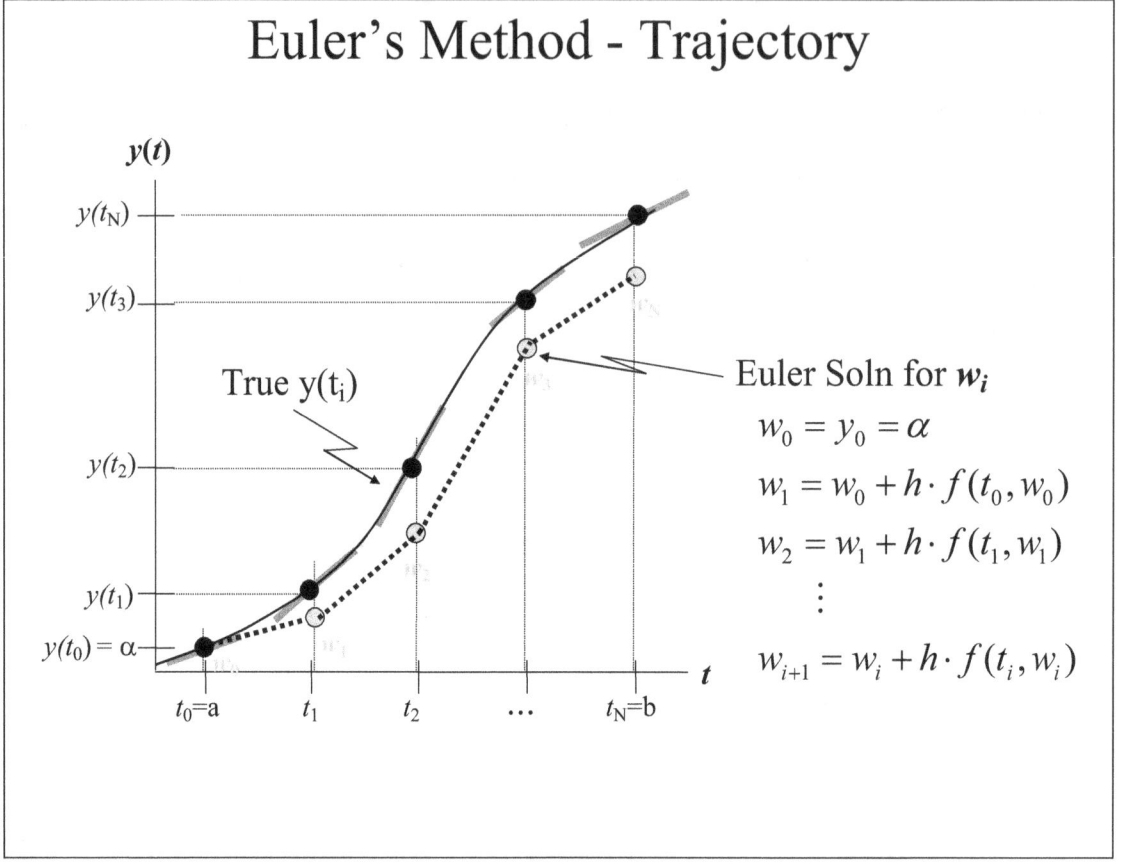

The figure shows the sequence of sampled y-values $\{y_0, y_1, y_2, ...\}$ as black dots and the true red slope segments $\arctan[f(t_i, y_i)]$ on the actual solution curve. The approximate solution generated by the Euler sequence of points $\{w_0, w_1, w_2, ...\}$ are shown as grey dots connected by straight line segments of length $f(t_i, w_i) \cdot h$. As we have already discussed, only for the initial point $t_0 = a$ do we have $w_0 = y_0$ and thereafter the "w_i"s no longer coincide with the "y_i"s as illustrated in the figure. This occurs for two reasons, namely, (i) at each time step we are continuing from two different points, i.e., (t_1, w_1) and (t_1, y_1), and (ii) the slopes (trajectory angles) $\theta_w = \arctan[f(t_1, w_1)]$ and $\theta_y = \arctan[f(t_1, y_1)]$ are different as well.

Several steps in the Euler solution sequence are shown and we see that the trajectory of points in the sequence is uniquely defined by the "start point" ($t=a$, $y=\alpha$) and the geometrical increments along the local Euler slopes: $w_1 = w_0 + f(t_0, w_0) h$, $w_2 = w_1 + f(t_1, w_1) h$, etc. .

We have already mentioned that the slope function is actually a surface $z(t,y) = f(t,y)$ above the t-y plane and the next two slides give visualizations of this 3^d surface. The local undulations of the surface determine whether the Euler solution actually tracks the true solution and the Lipschitz condition expressed in terms of $|f(t_i, y_i) - f(t_i, w_i)|$ becomes more intuitive.

2.1.2 Euler's Method – 3^d Representation & Slope Surface

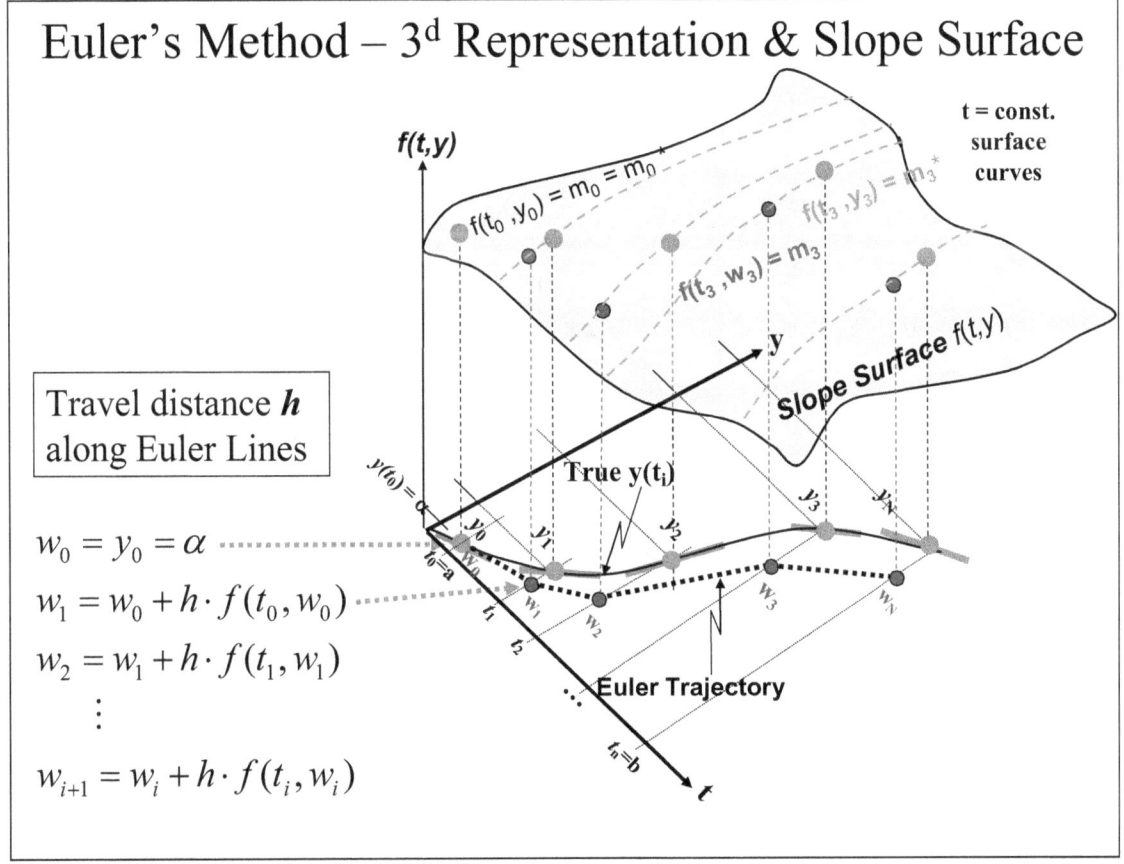

This 3^d representation of the Euler Method discussed in the last slide shows the slope surface f(t,y) above the t-y plane and the slope z-values corresponding to both the true and Euler trajectories. Clearly if we start these two trajectories at a common point and the approximate slope $m_1 = f(t_1, w_1)$ has a z-value that is close to the true slope $m_1^* = f(t_1, y_1)$, then the Euler trajectory will track the true solution curve. Singularities or very large changes in the slope for small changes in y-values will have a detrimental effect and may lead to instability.

2.1.3 Euler's Method - Perturbations

Euler's Method - Perturbations

$y' = t$; $y(0) = 0$; $t \in [0,1]$

$f(t,y) = t$
Slope Surface

$$y = \frac{t^2}{2}$$

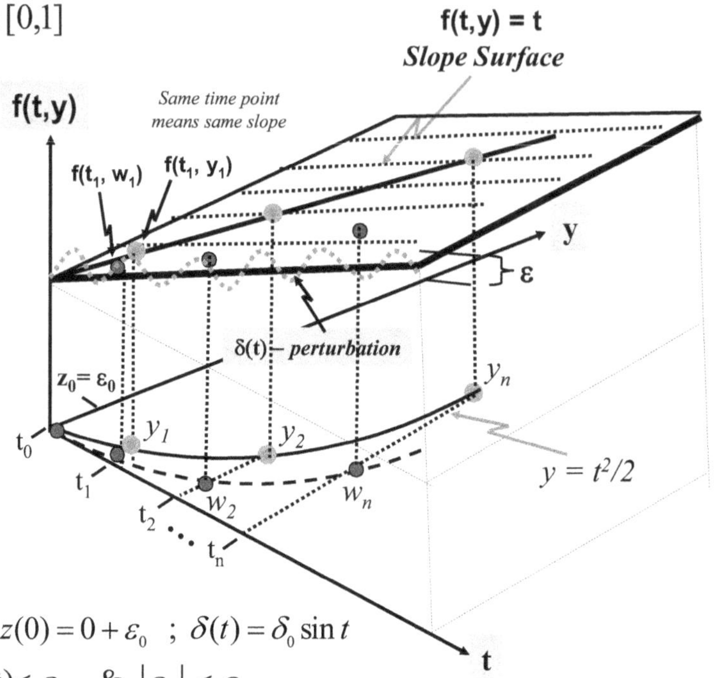

- Slope Surf indep of y;
- Only depends upon t
- Thus $f(t_1, w_1) = f(t_1, y_1)$, etc.,
- Correct slope at every mesh point
- Lifschitz Const. = 0 ($\partial f / \partial y = 0$)
- Perturbation

$z' = f(t,z) + \delta(t)$; $z(0) = 0 + \varepsilon_0$; $\delta(t) = \delta_0 \sin t$

$|y - z| \le k\varepsilon$ if $\delta(t) < \varepsilon$ & $|\varepsilon_0| < \varepsilon$

We seek a solution the DE $y' = f(t,y) = t$ over $t \in [0,1]$ with the IC $y(0)=0$. The slope surface has no dependence on y and is in fact a plane canted at 45° to the t-y plane as illustrated in the figure. The continuity and Lipschitz condition are clearly satisfied with a Lipschitz constant $\partial f(t,y)/\partial y = 0$ since $f(t,y)$ is independent of y; accordingly, the Euler Method has no problem tracking the true solution $y = t^2/2$ and the only consideration is whether this solution is stable to small perturbations.

We now add an undulating surface $\delta(t) = \delta_0 \sin(t)$ to the canted plane and start z at ε_0 and consider the perturbed IVP: $z' = f(t,z(t)) + \delta(t)$ with IC $z(0) = \varepsilon_0$. It is an easy matter to show by direct solution of the perturbed IVP that the quantity $|z(t) - y(t)| < k\varepsilon$, where $\varepsilon = \max(\varepsilon_0, \delta_0)$ and k is a fixed positive constant for all $t \in [0,1]$. Thus, the perturbed solution $z(t)$ is within $k\varepsilon$ of $y(t)$ for all t and hence the solution is stable. Note that stability of the perturbation example on the slide is easily verified by integrating the perturbed equation and putting a bound on $|y-z|$ as follows:

$$\int_{t=0}^{t} dz = \int_0^t (t + \delta_0 \sin t) dt \Rightarrow z(t) = z(0) + t^2/2 - \delta_0 \cos t = \varepsilon_0 + t^2/2 - \delta_0 \cos t$$

$$\therefore |y - z| = |t^2/2 - (\varepsilon_0 + t^2/2 - \delta_0 \cos t)|$$

$$= |-\varepsilon_0 + \delta_0 \cos t| \le |\varepsilon_0| + |\delta_0| \le 2 \cdot \max(\varepsilon_0, \delta_0) \le 2\varepsilon$$

2.2 Euler's Method - Error Analysis

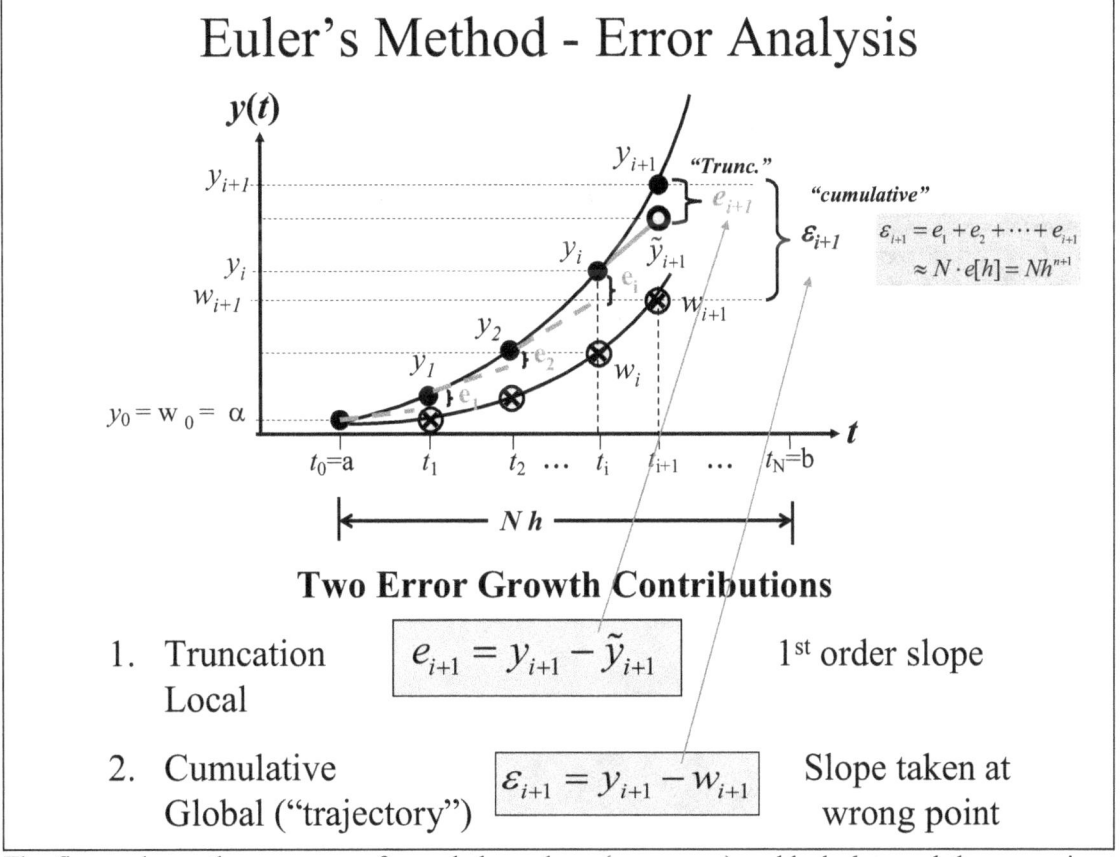

The figure shows the sequence of sampled y-values {$y_0, y_1, y_2, \ldots$} as black dots and the approximate solution generated by the Euler sequence of points {$w_0, w_1, w_2, \ldots$} shown as circled "x" s connected by straight line segments $f(t_i, w_i)h$. As we have already discussed, only for the initial point $t_0 = a$ do we have $w_0 = y_0 = \alpha$ and thereafter the "w_i"s no longer coincide with the "y_i"s as illustrated in the figure. The cumulative or "global truncation error at a given time t_{i+1} is the difference between the true and Euler values at that time, viz., $\varepsilon_{i+1} = y_{i+1} - w_{i+1}$ and is illustrated by the outer braces at the last point in the figure. Since the error produced by the Euler straight line segment from t_i to t_{i+1} results from both an error in the starting y-value (w_i) and in the slope itself $f(t_i, w_i)$ at that y-value, it is useful for analysis purposes to separate out the error associated with these two contributions.

Thus in the figure we show a y-value "y_{i+1}-tilde" at time t_{i+1} which would result if we actually started at the correct point (t_i, y_i), but used the Euler slope $f(t_i, y_i)$; the local truncation error is defined to be the difference $e_{i+1} = (y_{i+1} - $ "y_{i+1}-tilde") which is indicated by the smaller error brace at the end time t_{i+1}. The local truncation error is related to the Taylor approximation and is of key importance when considering higher order Taylor expansions; the contribution ("y_{i+1}-tilde" $- w_{i+1}$) results from an Euler step using the exact Euler slope $f(t_i, y_i)$, but from the wrong start point "w_i" (rather than "y_i").

We shall also see that when divided by the stepsize h the quantity $\tau[h] = e_{i+1}/h$ (also called "local truncation error") is used to test how well the difference equation method tracks the exact solution y_{i+1} to the DE. Taking the limit $h \to 0$ of $\tau[h] = \lim_{h \to 0}\{(y_{i+1} - y_i)/h - f(t_i, y_i)\}$ we see that if this quantity approaches zero we are essentially verifying that the **Euler slope $f(t_i, y_i)$ method is equivalent to the original DE**, viz., $y' = f(t_i, y_i)$ since $\lim_{h \to 0} (y_{i+1} - y_i)/h \to y'$ the derivative.

2.2.1 Euler & Higher Order Cumulative Error

Euler & Higher Order Cumulative Error

- **If** Single Step Error $e[h] = O(h^{n+1})$ n^{th} order Taylor
- **Then** Cum. Error $\varepsilon_N[h] = O(h^n)$ Down 1 order in h
- Crude approx

$$\varepsilon_N[h] \cong N \cdot e[h] = N \cdot h^{n+1} = (Nh) \cdot h^n = (b-a) \cdot h^n$$

Separating out the "local truncation error" as we did in the last slide allows us to state an important general relationship between the **local and global (cumulative) errors** for n^{th} order Taylor methods. The "rule of thumb" states that **if** we use an nth order Taylor method with **single step error** $e[h] = O(h^{n+1})$, **then** the cumulative error after N steps is down one order in h: $\varepsilon_N[h] = O(h^n)$. This can approximated by assuming that the cumulative effect of the local error is to simply sum up the N individual contributions with local truncation error $e[h]$ to yield N* $e[h]$ and then recognizing that one factor of h multiplies N to give the fixed interval (b-a) and there results $\varepsilon_N[h] = (b-a)\, h^n$ which is down one order. The next slide gives a more rigorous justification for this result in the Euler case and a theorem for the general case is stated later on.

2.2.2 Euler Convergence Theorem

Euler Convergence Theorem

- **1-Step Taylor:**
$$y_{i+1} \equiv y(t_{i+1}) = y(t_i) + y'(t_i) \cdot h + y''(\xi) \cdot \frac{h^2}{2}$$

- **Euler Approx.** $\quad = f(t_i, y_i)$
$$w_{i+1} = w_i + f(t_i, w_i) \cdot h$$

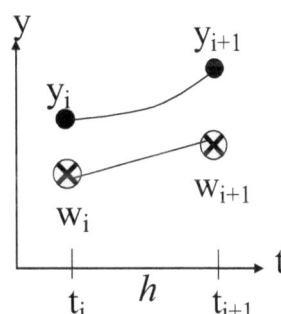

- **Cum. Error** (subst. y_{i+1}, w_{i+1})

$$\varepsilon_{i+1} = |y_{i+1} - w_{i+1}| \leq \underbrace{|y_i - w_i|}_{\equiv \varepsilon_i} + h \cdot \underbrace{|f(t_i, y_i) - f(t_i, w_i)|}_{\text{Lipschitz}: \leq L|y_i - w_i|} + \frac{h^2}{2} \cdot \underbrace{|y''(\xi_i)|}_{\leq M}$$

$$\varepsilon_{i+1} \leq \underbrace{(1 + hL)}_{\equiv s} \varepsilon_i + \underbrace{\frac{h^2}{2} M}_{\equiv r} \quad \text{or} \quad \boxed{\varepsilon_{i+1} = (1 + s)\varepsilon_i + r} \quad \text{First Order Difference Equation}$$

Solution is Linear in "h" at a Fixed Time t_{i+1}

$$t_{i+1} - a = (i+1) \cdot h = (i+1) \cdot \frac{b-a}{N}$$

$$\boxed{|y_{i+1} - w_{i+1}| \leq \frac{hM}{2L} \left[e^{L(t_{i+1} - a)} - 1 \right]}$$

The 1st order Taylor expansion of $y'(t)$ about $t = t_i$ has a local truncation error of $O(h^2)$ as given in the 1st equation; the associated Euler difference equation is given by the 2nd equation. The figure focuses on the step between times t_i and t_{i+1} and shows the true values y_i and y_{i+1} and the Euler values w_i and w_{i+1}. The cumulative error ε_{i+1} is obtained by subtracting the 1st two equations, taking the absolute value and recognizing that the Lipschitz condition means
$$| f(t_i, y_i) - f(t_i, w_i) | \leq L | y_i - w_i | ;$$
further defining max$\{y''(x)\} \leq M$ and the quantities $s = hL$ and $r = h^2 M/2$, we arrive at the difference equation
$$\varepsilon_{i+1} = (1+s) \varepsilon_i + r \quad \text{or} \quad \varepsilon_{i+1} - (1+s) \varepsilon_i = r \quad \text{with IC: } \varepsilon_0 = 0 \text{ (because for } i=0 \text{ we have } \varepsilon_0 = |y_0 - w_0| = 0 \text{)}$$
The constant solution to the full equation $\varepsilon_{i+1} = \varepsilon_i = -r/s$ is verified by direct substitution into the 2nd form, viz.,
$$(-r/s) - (1+s)(-r/s) = -r/s + r/s + r = r$$
Substitution of the trial solution $\varepsilon_i = \lambda^i$ into the homogeneous equation $\varepsilon_{i+1} - (1+s) \varepsilon_i = 0$ yields the polynomial
$$\lambda^{i+1} - (1+s) \lambda^i = 0 \quad \Rightarrow \quad \lambda = (1+s) \text{ is the non-trivial root}$$
Thus by superposition, the general solution is the sum of these two solutions
$$\varepsilon_i = C\lambda^i - r/s = C(1+s)^i - r/s$$
Applying the initial condition by setting $i=0$ we have
$$\varepsilon_0 = C(1+s)^0 - r/s \quad \Rightarrow \quad C = r/s$$
and there results
$$\varepsilon_i = (r/s)(1+s)^i - r/s = (r/s) [(1+s)^i - 1] = (h^2 M/2)/(hL) [(1+hL)^i - 1]$$
Noting that to 1st order the term $(1+hL)^i \sim (e^{hL})^i$ and also that the time is defined as $t_i = a + ih$ or $i = (t_i - a)/h$, we have
$$\varepsilon_i = (hM)/(2L) [e^{ihL} - 1] = (hM)/(2L) [\exp(L(t_i - a)) - 1]$$
where we have used $ihL = L(t_i - a)$ in the second equality. This last equation shows that the cumulative error is linear in h, i.e., $\varepsilon_i \sim O(h)$, which is down one order from the single step or local truncation error of $O(h^2)$.

2.2.3 Euler's Method: Round Off vs. Truncation Error

Euler's Method: Round Off vs. Truncation Error

$$w_0 = \alpha$$
$$w_{i+1} = w_i + h \cdot f(t_i, w_i)$$

$$\rightarrow \begin{cases} u_0 = \hat{\alpha} \\ u_{i+1} = u_i + h \cdot f(t_i, u_i) + \delta_i \end{cases}$$

$$\Delta\alpha = \hat{\alpha} - \alpha \quad ; \quad |\delta_i| < \delta$$

$$|y_i - u_i| \leq \frac{1}{L}\left(\underbrace{\frac{Mh}{2}}_{Trunc} + \underbrace{\frac{\delta}{h}}_{RO}\right)\left[e^{L(t_i - a)} - 1\right] + (\Delta\alpha)e^{L(t_i - a)}$$

- Error bound is *unstable* as $h \rightarrow 0$
- *Optimal* for $h = \sqrt{\dfrac{2\delta}{M}}$

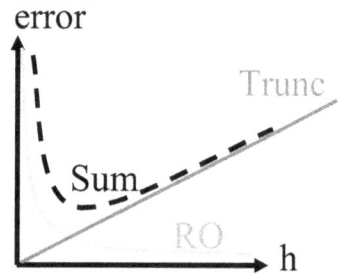

In order to include round off error we need to solve a new difference equation for which the initial condition is changed from α to $\alpha+\Delta\alpha$ and the slope evaluation has a RO error δ_i different for each step. Now forming the absolute difference $\varepsilon_i = |y_i - u_i|$ and performing the same analysis as before we obtain the boxed equation for the total error ε_i at time step t_i. This looks similar to our previous result except that in addition to the truncation term found previously, there is a term containing the maximum of the RO errors δ_i; moreover, there is now a second term **independent of h** which contains the RO error $\Delta\alpha$.

As we have seen previously in computing numerical derivatives there is a trade-off between the truncation error (red straight line) which varies as $\sim h$ and the RO error (green hyperbola) which varies as $1/h$. The familiar plot of these two errors shows the RO error becoming unbounded for small h and the truncation error decreasing for small h and their sum (dashed curve) yielding an optimal choice of stepsize at its minimum. Thus the solution becomes unstable if we let h approach zero because we lose all significant digits by virtue of the increasing RO error.

2.3 Need for Higher Order Methods

Need for Higher Order Methods

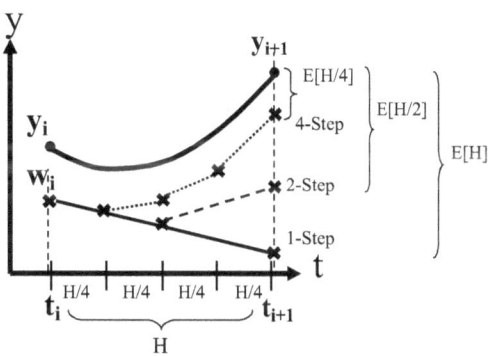

- For given Trunc. Error fewer sub-steps *needed* with higher order method
- *This means* less RO error (fewer ops)

Truncation Error for r substeps h=H/r

- Per step local Trunc. Error $e[h] \sim Ch^{n+1}$
- Cum. Error after r substeps $E[h] \sim Ch^n$
  (down one order)

- Thus $\boxed{E[h] \cong C \cdot h^n = C\left(\dfrac{H}{r}\right)^n = \dfrac{1}{r^n} \cdot E[H]}$

- Reduce Trunc error by factor of 100 (say)
- Means $r^n = 100 \rightarrow r = (100)^{1/n}$

Single step H vs. r substeps h=H/r

Method	Order n	#substeps r	Equiv. Num h-steps
Euler	1	100	100 *(H/100)
Mod. Euler	2	10	10 *(H/10)
RK4	4	$(100)^{1/4}$= 3	3 *(H/3)

#Steps for same reduction in truncation error

The need for higher order methods in solving our IVP is essentially the same here as it was for performing numerical quadratures on integrals, *i.e.*, increasing the order of the method decreases the *single step truncation error* and by increasing the number of steps for *lower order methods* we can compare different order methods with the same effective truncation error. Larger step sizes h in higher order methods decreases RO error and requires fewer total operations to compute the value at the output step H. The figure compares results for one integration step between t_i and t_{i+1} for a single large step H with that for 4 smaller steps of size h=H/4. It is clear that if we use the same method (say Euler) then the 4 smaller steps track the true solution more closely and has smaller truncation error than the one large step; however this implies 4 times as many arithmetic operations and hence increased RO error. So we must trade off RO and truncation error.

The trade off can be analyzed by using our rule of thumb (proven for the Euler Method) that the cumulative error at t_{i+1} is down one order. Thus, the local truncation error $e[h] \sim Ch^{n+1}$ results in a cumulative error
$$\varepsilon[h] = Ch^n = C(H/r)^n = (1/r^n)\varepsilon[H],$$
where the smaller step reduces H by a factor "r" : h=H/r .

If we wish to reduce the truncation error by, say, a factor of 100, then $r^n = 100$ and the table gives the results for several methods, namely, order n = 1(Euler) , n = 2(Modified Euler), and n = 4 (RK4) .

The table shows that in order to improve the truncation error by a factor of 100 using the Euler method requires us to make 100 sub-steps of size H/100, while it only requires 10 of step size H/10 for modified Euler (see Slide# 2-16), and 3 of step size H/3 for RK4. Given that the truncation error is the same for all three methods, we see that RK4 is the preferred method since it has fewer operations and hence smaller RO error.

Self-Starting Single Step Methods

2.3.1 Method-Sampled Slopes

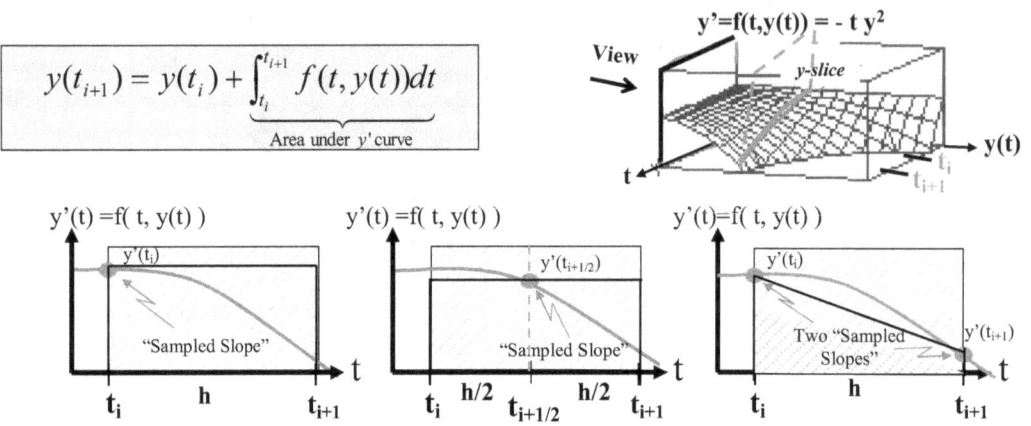

Intuitive Generalization of Euler's Method—*Sampled Slopes*

$$y(t_{i+1}) = y(t_i) + \underbrace{\int_{t_i}^{t_{i+1}} f(t, y(t))dt}_{\text{Area under } y' \text{ curve}}$$

$y' = f(t, y(t)) = -t y^2$

	Euler	Midpoint	Trapezoidal
Intgrl Area	Area = h y'(t_i)	Area = h y'($t_{i+1/2}$)	Area = ½ h [y'(t_i) + y'(t_{i+1})]
Est. Slopes	= h f (t_i, w_i)	= h f ($t_{i+1/2}$, w_i + ½ h f_i)	= ½ h [f (t_i, w_i) + f (t_{i+1}, w_i + h f_i)]
Param	$\alpha = 0$	$\alpha = ½$	$\alpha = 0, 1$
Notes	Estimate Slope function f (t_i, w_i) at time: $t_{i+\alpha} = t_i + \alpha h$ and at y-value obtained by traveling α h along the "Euler Line" $w_{i+\alpha} = w_i + \alpha h f_i$		

The boxed equation gives a formal solution to the DE over the interval [t_i, t_{i+1}] which expresses the solution y(t_{i+1}) at t=t_{i+1} as the sum of its previous value y(t_i) and the integral of f(t,y(t)) over the interval [t_i, t_{i+1}]. If we "knew" y(t) as a function of "t" we could integrate directly, but in that case there would be no need to perform the integral since its sole purpose is to produce a new value of the solution at y(t_{i+1}) which is "known". As illustrated in the 3-dim plot of the slope surface f(t,y) (top figure), a cut through this surface at y = c = constant yields a unique curve in the y'- t plane. Whereas y(t) is unknown, y'(t) for a fixed value of y=c is a known function say f(t,y) =-ty² =-tc² and hence the boxed integral can be approximated by computing the "area" under the y'(t) curve as h f(t,y). Thus sketching the function y'(t) in the t-y' plane allows us to consider solving the problem in the context of estimating the area under that curve. The examples in the lower figures and table illustrate how the Euler, Midpoint, and Trapezoidal methods may be viewed as area calculations. Since we know t_i and w_i and thus the slope at f(t_i, w_i), we can approximate slopes at other points by defining $t_{i+\alpha} = t_i + \alpha h$ and setting $w_{i+\alpha} = w_i + \alpha h$ f(t_i, w_i) (Euler line).

The Euler case (col#1) approximates the area under the curve as the shaded rectangle with height y'(t_i) and length "h" and estimates y'(t_i) by evaluating the slope function at the point (t_i, w_i), so the Area = h f (t_i, w_i)

The Midpoint case (col#2), as the name suggests, uses the slope function at the midpoint ($t_{i+½}$, $w_{i+½}$), where $t_{i+½} = t_i + ½$ h, $w_{i+½} = w_i + (h/2)$ f(t_i, w_i), and the rectangle is Area = h f ($t_{i+½}$, $w_{i+½}$).

The Trapezoidal case (col#3), evaluates the slope function at both end points (t_i, w_i) and (t_{i+1}, w_{i+1}) and computes the trapezoidal area with height h and bases f (t_i, w_i) and f (t_{i+1}, w_{i+1}) as

$$\text{Area} = ½ \, h[f (t_i, w_i) + f (t_{i+1}, w_{i+1})]$$

In general we might choose to evaluate slopes at arbitrary points within or outside the interval [t_i, t_{i+1}]. by introducing a parameter α and defining $t_{i+\alpha} = t_i + \alpha$ h and $w_{i+\alpha} = w_i + \alpha$ h f(t_i, w_i) and the three cases would correspond to $\alpha = 0$, $\alpha = ½$, and the pair $\alpha = 0,1$ respectively.

2.3.2 Sampled Slope Formalism

Sampled Slope Formalism

Formal Solution (one step)

$$y' = f(t, y(t)) \rightarrow \int_{t_i}^{t_{i+1}} dy = \int_{t_i}^{t_{i+1}} f(t, y(t)) dt$$

$$\underbrace{y(t_{i+1})}_{\equiv y_{i+1}} = \underbrace{y(t_i)}_{\equiv y_i} + \underbrace{\int_{t_i}^{t_{i+1}} f(t, y(t)) dt}_{\equiv h\phi}$$

Approx. Solution:

$$y_{i+1} \cong w_{i+1} = w_i + h\phi \; ; \; h\phi \approx \int_{t_i}^{t_{i+1}} f(t, y(t)) dt$$

Weighted sum of slopes:

$$h\phi = h \sum_i a_i \cdot m_i \; ; \; m_i \equiv f(t_{i+\alpha}, w_{i+\alpha})$$

Parameter α:

$$t_{i+\alpha} \equiv t_i + \alpha h$$

$$f_i \equiv f(t_i, w_i)$$

$$w_{i+\alpha} = w_i + \underbrace{\alpha h f_i}_{\text{Euler Line}}$$

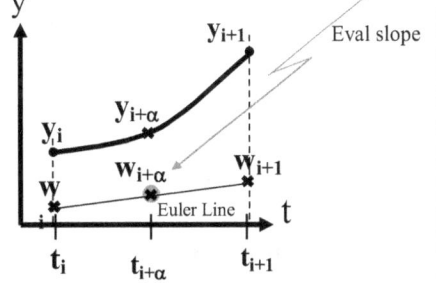

The formal solution to the DE over the interval [t_i, t_{i+1}] is re-written as the difference equation

$$w_{i+1} = w_i + h\phi$$

where ϕ is simply a weighted sum of slopes = $f(t_{i+\alpha}, w_{i+\alpha})$. As the figure illustrates the slope(s) m_i are evaluated at one or more points in the t-y plane "$t_{i+\alpha}$, $w_{i+\alpha}$" and a weighted sum specific to the algorithm approximates the area and hence gives the increment that must be added to w_i to get w_{i+1}.
It must be emphasized that this Area is in the t-y'plane (figures in previous slide), not in the t-y plane shown in this slide. The Runge-Kutta solution of various orders are most easily understood by finding appropriate "weighted sum of slopes" that can be made to match a Taylor polynomial for the given order n=2,3,4. .
The RK2 case (n=2) derivation turns out to be quite easy and instructive, while the RK3 (n=3) derivation is doable with some effort. However, the "Runge-Kutta formula" RK4 (n=4) is a very tedious and difficult result to derive.

2.3.3 Higher Order Taylor Method

Higher Order Taylor Method

Taylor Series Expansion:	$y(t_i + h) = y(t_i) + h \cdot y'(t_i) + \dfrac{h^2}{2!} \cdot y''(t_i) + \dfrac{h^3}{3!} \cdot y'''(t_i) + \cdots + \dfrac{h^n}{n!} \cdot y^{(n)}(t_i) + R_{n+1}(\xi)$ $R_{n+1}(\xi) \equiv y^{(n+1)}(\xi) \cdot \dfrac{h^{n+1}}{(n+1)!}\ ;\ \xi \in [t_i, t_{i+1}]$
Successive Derivatives of DE:	$y'(t) = f(t, y(t))\ ;\ y''(t) = \dfrac{d}{dt} f(t, y(t))\ ;\ \cdots\ ;\ y^{(n)}(t) = \dfrac{d^{n-1}}{dt^{n-1}} f(t, y(t))$
n^{th} order Taylor Approx:	$y(t_i + h) = y(t_i) + h \cdot \underbrace{\left[f(t, y(t)) + \dfrac{h}{2!} \dfrac{d}{dt} f(t, y(t)) + \dfrac{h^2}{3!} \dfrac{d^2}{dt^2} f(t, y(t)) + \cdots + \dfrac{h^{n-1}}{n!} \dfrac{d^{n-1}}{dt^{n-1}} f(t, y(t)) \right]}_{\phi_{Taylor}}\bigg|_{t=t_i} + R_{n+1}(\xi)$
Associated n^{th} order difference eqn:	$w_0 = \alpha$ (I.C. or Start Value) $w_{i+1} = w_i + h \cdot \underbrace{\left[f_i + \dfrac{h}{2!} \cdot f_i' + \dfrac{h^2}{3!} \cdot f_i'' + \cdots + \dfrac{h^{n-1}}{n!} \cdot f_i^{(n-1)} \right]}_{\equiv \phi_{Taylor}}$
Note: Taylor Method Requires Derivatives!! ⟹	$f_i = f(t_i, w_i)\ ;\ f_i^{(n-1)} \equiv \dfrac{d^{n-1}}{dt^{n-1}} f(t, y(t)) \bigg|_{\substack{t=t_i \\ y(t)=w_i}}$

The nth order Taylor expansion of $y(t_i + h)$ about $t = t_i$ has $(n+1)$ terms up to h^n and a local truncation error $\varepsilon \sim O(h^{n+1})$ as given by the 1st equation on the slide. We can formally compute the necessary derivatives $y'(t), y''(t), \ldots, y^{(n)}(t)$ by repeatedly differentiating the DE $y'(t) = f(t,y(t))$, to obtain $y''(t) = d/dt[f(t,y(t))]$, ...etc. and substitution yields the nth order Taylor approximation with error term as shown. The associated difference equation is obtained by dropping the error term and replacing the "y_i"s with "w_i"s as displayed in the shaded box.

The "total" derivatives with respect to time t denote $f_i', f_i'', \ldots, f_i^{(n-1)}$ are computed from the explicit functional form of $f(t,y(t))$ by using the DE to evaluate $d/dt\,[y(t)]$ as $f(t,y)$ whenever it occurs. For example, if $f(t,y(t)) = -\sin(t)/y$, then the first few derivatives are computed as follows:

$f = f(t,y(t)) = -\sin(t)/y \Rightarrow -\sin(t_i)/w_i$

$f' = d/dt\,[-\sin(t)/y] = -\cos(t)/y - \sin(t)(-1/y^2)\,y' = -\cos(t)/y - \sin(t)(-1/y^2)[-\sin(t)/y]$
$\quad = [-\cos(t)/y - \sin^2(t)/y^3] \Rightarrow -\cos(t_i)/w_i - \sin^2(t_i)/w_i^3$

$f'' = d/dt[-\cos(t)/y - \sin^2(t)/y^3] = +\sin(t)/y + \cos(t)/y^2 \cdot (-\sin(t)/y) - 2\sin(t)\cos(t)/y^3 - \sin^2(t)(-3y^{-4})\cdot(-\sin(t)/y)$
$\quad = +\sin(t)/y - 3\sin(t)\cos(t)/y^3 - 3\sin^3(t)/y^5 \Rightarrow$
$\quad = \sin(t_i)/w_i - 3\sin(t_i)\cos(t_i)/w_i^3 - 3\sin^3(t_i)/w_i^5$

According to the boxed equation the 3rd order Taylor Method for this example is
$w_{i+1} = w_i + h\,[\,-\sin(t_i)/w_i + (h/2)\,\{-\cos(t_i)/w_i - \sin^2(t_i)/w_i^3\}$
$\qquad\qquad + h^2/6\,\{\sin(t_i)/w_i - 3\sin(t_i)\cos(t_i)/w_i^3 - 3\sin^3(t_i)/w_i^5\}\,]$

Step-by-step generation of the sequence for the IVP $y' = -\sin(t)/y$ for $t \in [\pi/2, 5\pi/2]$ with the IC $y(t=\pi/2) = 2$ is straightforward once we choose our time step increment $h = [5\pi/2 - \pi/2]/10 = 0.2\pi$ so the times are given by $t_i = \pi/2 + i\,h = [\pi/2 + 0.2\pi\,i\,]$.

For $i=0$ we have $t_0 = \pi/2$ and $w_0 = y(\pi/2) = 2$; we then compute the next "w-value" as
$\quad w_1 = 2 + 0.2\pi\,[\,-\sin(\pi/2)/2 + (.2\pi/2)\,\{-\cos(\pi/2)/2 - \sin^2(\pi/2)/2^3\}$
$\qquad\qquad + ((.2\pi)^2/6)\{\sin(\pi/2)/2 - 3\sin(\pi/2)\cos(\pi/2)/2^3 - 3\sin^3(\pi/2)/2^5\}\,] = 2.306$

Similarly, $i=1$ yields $t_1 = \pi/2 + 0.2\pi*1 = .7\pi$ and $w_2 = 2.306 + 0.2\pi\cdot\phi[t_1 = .7\pi, w_1 = 2.306\,]$, etc.

2.3.4 Example: Taylor's Method n =2,3

Example: Taylor's Method n =2,3

IVP: $y' = f(t, y(t)) = -y + t + 1 \; ; \; y(0) = 1 \; ; \; t \in [0,1]$

Derivs: $f' \equiv \dfrac{d}{dt} f(t, y(t)) = \dfrac{\partial}{\partial t} f(t, y(t)) + \underbrace{\dfrac{dy}{dt}}_{\equiv y'} \dfrac{\partial}{\partial y} f(t, y(t)) = 1 + y'(-1) = 1 - (-y + t + 1) = y - t$

Note: Partials wrt the t and y

$f'' \equiv \dfrac{d}{dt} f' = \dfrac{\partial}{\partial t}(y - t) + y' \dfrac{\partial}{\partial y}(y - t) = -1 + y'(+1) = -1 + (-y + t + 1) = -y + t$

Results:

$n = 2 \quad \begin{cases} w_0 = \alpha = 1 \\ w_{i+1} = w_i + h\left[(-w_i + t_i + 1) + \dfrac{h}{2}(w_i - t_i)\right] = w_i + h\left[1 + \left(\dfrac{h}{2} - 1\right)(w_i - t_i)\right] \end{cases}$

$n = 3 \quad \begin{cases} w_0 = \alpha = 1 \\ w_{i+1} = w_i + h\left[(-w_i + t_i + 1) + \dfrac{h}{2}(w_i - t_i) + \dfrac{h^2}{6}(-w_i + t_i)\right] = w_i + h\left[1 + \left(-\dfrac{h^2}{6} + \dfrac{h}{2} - 1\right)(w_i - t_i)\right] \end{cases}$

The Taylor Methods of order n=2,3 for the IVP: $y' = f(t,y(t) = -y+t+1$; $y(0) = 1$ for $t \in [0,1]$ are obtained by computing the required derivatives and substituting into the boxed equation of the previous slide. The "total" derivatives with respect to time t denoted f_i', f_i'', ..., $f_i^{(n-1)}$ are computed from the explicit functional form of f(t,y(t)) by using the DE to evaluate d/dt [y(t)] as f(t,y) whenever it occurs. When we have an explicit expression for f(t,y) the process of computing the derivatives is straight forward as shown in the slide. However, it should be noted that the total derivative of f(t,y) is formally a derivative of a function of two variables, namely "t" and "y" so the total derivative d/dt of f(t,y(t)) is formally evaluated as the derivative of a function of two variables using the chain rule on y(t) as follows:

$$d/dt[f(t,y)] = \partial/\partial t \, [f(t,y)] + (\partial y/\partial t) \partial/\partial y \, [f(t,y)]$$
$$= \partial/\partial t \, [f(t,y)] + (dy/dt)\partial/\partial y \, [f(t,y)]$$
$$= \{\partial/\partial t + f(t,y)\partial/\partial y \} \, [f(t,y)]$$

The second equality above results from the fact that y is only a function of "t" and thus the partial $\partial y/\partial t$ is just the total derivative dy/dt. Making this substitution from the DE dy/dt = f(t,y) yields the final form in which the total derivative is expressed as the operator $\{\partial/\partial t + f(t,y)\partial/\partial y \}$ acting on the function of two variables f(t,y). In this "operator" form the 2nd derivative is

$$d^2/dt^2 = \{\partial/\partial t + f(t,y)\partial/\partial y \}^2$$

and the nth derivative is

$$d^n/dt^n = \{\partial/\partial t + f(t,y)\partial/\partial y \}^n.$$

Although this formal operator development is not needed for the current example, it will be crucial in understanding the derivation of the Runge-Kutta RK2 Method on the next slide.

Self-Starting Single Step Methods

2.4 2nd Order Runge-Kutta Derivation

2nd Order Runge-Kutta Derivation

2nd Order Taylor: $\quad y_{i+1} = y_i + h \cdot y'_i + \dfrac{h^2}{2} \cdot y''_i + O(h^3)$

Time Derivs. $\quad y'_i = f(t_i, y_i)$

$$y''_i = \dfrac{d}{dt} f(t_i, y_i) = \dfrac{\partial}{\partial t} f(t_i, y_i) + \dfrac{dy_i}{dt} \cdot \dfrac{\partial}{\partial y_i} f(t_i, y_i)$$

$$= f_t + f_y \cdot y' = f_t + f \cdot f_y$$

Total Time Deriv. Operator:
$$\dfrac{d}{dt} \equiv \dfrac{\partial}{\partial t} + f \dfrac{\partial}{\partial y}$$

Subs Derivs into Taylor: $\quad y_{i+1} = y_i + \left[h \cdot f(t_i, y_i) + \dfrac{h^2}{2} \cdot (f_t + f_y f(t_i, y_i)) \right] + O(h^3)$

Assoc Difference Eqn: $\quad w_{i+1} = w_i + h \cdot \left[f(t_i, w_i) + \dfrac{h}{2} \cdot (f_t + f_y f(t_i, w_i)) \right] \quad\quad \phi^{(2)}_{Taylor} = f + \dfrac{h}{2} f_t + \dfrac{h}{2} f_y f$

Generalize Slope Fcn: $\quad \phi^{(2)}_{Slope} = a_1 f(t_i, w_i) + a_2 f(\underbrace{t_i + \alpha h}_{=\Delta t}, \underbrace{w_i + \beta h f}_{=\Delta y})$

$$\phi^{(2)}_{Slope} = a_1 f(t_i, w_i) + a_2 \underbrace{\left[f(t_i, w_i) + \dfrac{\partial f}{\partial t} \Delta t + \dfrac{\partial f}{\partial y} \Delta y + O(\Delta t^2, \Delta y^2) \right]}_{2^d \text{ Taylor Expansion abt } (t_i, w_i)}$$

Compare

$$\phi^{(2)}_{Slope} \cong (a_1 + a_2) f + a_2 f_t \cdot \alpha h + a_2 f_y \cdot \beta h f$$

Equate $\phi^{(2)}_{Taylor} = \phi^{(2)}_{Slope}$ **Term-by-term:**

$$f : 1 = (a_1 + a_2) \Rightarrow \boxed{a_1 = 1 - a_2}$$
$$f_t : \dfrac{h}{2} = a_2 \alpha h$$
$$f f_y : \dfrac{h}{2} = a_2 \beta h \Biggr\} \Rightarrow \alpha = \beta = \dfrac{1}{2 a_2}$$

Arbitrary Parameter "a_2" yields RK2 Family of Solns

Lets consider the 2nd order Taylor Method and formally expand it in terms of functional derivatives. We found that the total derivative of the slope function f(t,y) for the 1st order DE could be expressed as an operator and we now re-write this expression in a simpler notation:

$$d/dt[f(t,y)] = \{\partial/\partial t + f(t,y)\partial/\partial y\} [f(t,y)] = f_t + f f_y$$

where the partials are taken as subscripts $f_t = \partial/\partial t [f(t,y)]$ and $f_y = \partial/\partial y [f(t,y)]$. Upon substitution into the 2nd order Taylor Method we obtain

$$y_{i+1} = y_i + h [f(t_i, y_i) + \tfrac{1}{2} h (f_t + f_y f(t_i, y_i))] + O(h^3)$$

with all terms on the RHS evaluated at the point $(t,y) = (t_i, y_i)$. The associated difference equation (DfE) is as usual obtained by dropping the error term and setting $y_i = w_i$, viz.,

$$w_{i+1} = w_i + h[f(t_i, y_i) + \tfrac{1}{2} h (f_t + f_y f(t_i, w_i))] \quad \text{or} \quad w_{i+1} = w_i + h \phi^{(2)}_{Taylor} \quad \text{with}$$

$$\phi^{(2)}_{Taylor}(t_i, w_i) = [f(t_i, w_i) + \tfrac{1}{2} h (f_t + f_y f(t_i, w_i))] = f + (\tfrac{1}{2} h) f_t + (\tfrac{1}{2} h) f f_y$$

Now consider the "sampled slope" formalism previously discussed when generalizing Euler's Method and write it down in terms of two weighting coefficients a_1 and a_2 and two additional parameters α and β used to specify the slope function evaluation points as follows

$$\phi^{(2)}_{Slope} = a_1 f(t_i, w_i) + a_2 f(t_i + \alpha h, w_i + \beta h f)$$

The specific form is arrived at by trial and error, but the trick here is to choose a form, which when expanded as a 2^d Taylor polynomial, matches the coefficients of the three distinct terms f, f_t, and $f f_y$ in $\phi^{(2)}_{Taylor}$. The general form of the 2^d Taylor expansion (to 1st order) is applied to $\phi^{(2)}_{Slope}(t_i + \Delta t, w_i + \Delta w)$ with $\Delta t = \alpha h$ and $\Delta y = \beta h f$ to yield

$$\phi^{(2)}_{Slope} = a_1 f + a_2 \{ f + f_t (\alpha h) + f_y (\beta h f) \} = (a_1 + a_2) f + (a_2 \alpha h) f_t + (a_2 \beta h) f f_y$$

Equating $\phi^{(2)}_{Slope}$ to $\phi^{(2)}_{Taylor}$ term-by-term yields three equations yields $a_1 = 1 - a_2$ and $\alpha = \beta = 1/(2 a_2)$, where a_2 is a **free parameter** whose choices generate a **"family"** of RK2 solutions.

2.4.1 RK2 Family of Solutions

RK2 Family of Solutions

- RK2 Family of Solutions Parametric in a_2

$$w_0 = \alpha \quad ; \quad w_{i+1} = w_i + h\phi_i$$

$$\phi_i \equiv (1-a_2)f(t_i, w_i) + a_2 f(t_i + \frac{h}{2a_2}, w_i + \frac{h}{2a_2}f(t_i, w_i))$$

a_2	$a_1 = 1 - a_2$	$\alpha = \beta = 1/(2a_2)$	Name	RK2 Recursion Equation
1	0	½	Midpoint Method	$w_{i+1} = w_i + h\, f(t_i + \frac{h}{2}, w_i + \frac{h}{2} \cdot f)$
½	½	1	Modified Euler (Trapezoidal)	$w_{i+1} = w_i + \frac{h}{2}[f(t_i, w_i) + f(t_i + h, w_i + hf)]$
¾	¼	2/3	Huen's Method	$w_{i+1} = w_i + \frac{h}{4}[f(t_i, w_i) + 3f(t_i + \frac{2}{3}h, w_i + \frac{2}{3}h \cdot f)]$
0	1	$a_2 = 0$ not applicable	Euler's Method (1st order)	$w_{i+1} = w_i + h \cdot f(t_i, w_i)$

- Euler is 1st order Method; not a member of RK2 Family
- Other Choices for a_2 are possible
- Trapezoidal Method is more commonly called Modified Euler Method
- Note: Runge-Kutta Methods do NOT Require Derivatives!!

The family of RK2 Methods in Sampled Slope form is shown explicitly in the boxed equation with the free parameter a_2 left unspecified. It must be emphasized that this solution is equivalent to (matches) the Taylor Method of order 2, but it is **no longer necessary to compute the analytic derivatives** of the slope function for each new problem, because the derivatives are **implicit in the sampled slope form itself.** That is, the **derivative computation for all slope functions** has been done up front in forming the equivalence of the two methods.

The table shows several choices of the free parameter a_2 (col#1), the specific parameter values (col #2, col#3) and the common names (col#4) for these members of this RK2 family. Note that there are infinitely many other choices for a_2 but we have only listed the common ones. All these methods are equivalent to the 2nd order Taylor Method in terms of truncation error and again do not require the computation of derivatives. Also note that the Trapezoidal Method discussed in the Sampled Slope formalism is more commonly referred to as the "modified Euler method" and is a member of the RK2 family.

Finally, although the Euler solution results from the general RK2 formulation, it is a degenerate case in which the 2nd slope multiplier $a_2 = 0$ and hence is a 1st order method, not a 2nd order method as are all members of the RK2 family. On the other hand, the Midpoint method has $a_1 = 0$, but the second multiplier a_2 is non-zero and the partials f_t and f_y of the 2nd order slope expansion $\phi^{(2)}_{Slope}$ and thus makes the method second order.

2.4.2 RK2 Example

RK2 Example

IVP: $y' = y$ $y(0) = 1$ **Soln**: $y = e^t$

RK2 Soln: $w_0 = 1$; $w_{i+1} = w_i + h\phi_i$

$$\phi_i \equiv (1-a_2)\underbrace{f(t_i, w_i)}_{=w_i} + a_2 \underbrace{f(t_i + \frac{h}{2a_2},\ w_i + \frac{h}{2a_2} f(t_i, w_i))}_{=w_i + \frac{h}{2a_2} w_i}$$

$$w_{i+1} = w_i + h\left[(1-a_2)w_i + a_2\left(w_i + \frac{h}{2a_2} w_i\right)\right] = w_i\left[1 + h + \frac{h^2}{2}\right]$$

Note: a_2 term cancels out.

Sum Series:

$$w_{i+1} = w_i\left[1 + h + \frac{h^2}{2}\right]$$

$i = 0$ $w_1 = w_0 \cdot \left[1 + h + \frac{h^2}{2}\right]$

$i = 1$ $w_2 = w_1 \cdot \left[1 + h + \frac{h^2}{2}\right]$

$i = 2$ $w_3 = w_2 \cdot \left[1 + h + \frac{h^2}{2}\right]$

$\vdots$

$i = k-1$ $w_k = w_{k-1} \cdot \left[1 + h + \frac{h^2}{2}\right]$

$$\Rightarrow w_1 w_2 w_3 \cdots w_{k-1} w_k = w_0 w_1 w_2 \cdots w_{k-1}\left[1 + h + \frac{h^2}{2}\right]^k$$

$$w_k = w_0\left[1 + h + \frac{h^2}{2}\right]^k \approx w_0\left[e^h\right]^k$$

Matches Exact Soln to 2nd order in h

$$= w_0 e^{hk} = 1 \cdot e^t$$

Here is a very trivial example where f(t,w) does not depend upon "t" { y'=y ; y(0)=1} whose solution is y = e^t. This IVP is interesting because we can analytically show that the whole family of RK2 methods gives the exact solution to 2nd order independent of our choice of the free parameter a_2. Thus substituting f(t,w) =w onto the general form of the RK2 method we easily find that the a_2-term cancels out to yield the solution
$w_{i+1} = w_i [1 + h + h^2/2]$
as detailed in the slide. This solution represents a simple 1st order recursion relation between successive terms w_i and w_{i+1}; writing out the recursion for i=0,1,2, ..., k-1 and multiplying the resulting equations allows cancellations of "w"-terms on the left and right side leaving
$w_k = w_0[1+h+h^2/2]^k \approx w_0[e^h]^k = w_0 e^{hk} = e^{t_k}$
where the "h"terms in the first square bracket are equivalent to the exponential e^h expanded to 2nd order and the term "hk" is simply the time t_k.

2.5 RK4 Derivation

RK4 Derivation

Slope Function (4th order):

$$w_{i+1} = w_i + h\phi_i^{(4)} \quad ; \quad h\phi_i^{(4)} = \sum_{j=1}^{4} a_j k_j$$

$$h\phi_i^{(4)} = a_1 \underbrace{hf(t_i, w_i)}_{\equiv k_1} + a_2 \underbrace{hf(t_i + \alpha_1 h, \ w_i + \beta_1 k_1)}_{\equiv k_2} + a_3 \underbrace{hf(t_i + \alpha_2 h, \ w_i + \gamma_1 k_1 + \gamma_2 k_2)}_{\equiv k_3}$$

$$+ a_4 \underbrace{hf(t_i + \alpha_3 h, \ w_i + \delta_1 k_1 + \delta_2 k_2 + \delta_3 k_3)}_{\equiv k_4}$$

2d Taylor Expansion of $\phi^{(4)}{}_{Slope}$:

$$k_1 = hf(t_i, w_i)$$

$$k_2 = hf(t_i + \underbrace{\alpha_1 h}_{\Delta t}, \ w_i + \underbrace{\beta_1 k_1}_{\Delta y})$$

$$= h\left\{ f(t_i, w_i) + \left[f_t \cdot (\alpha_1 h) + f_y \cdot (\beta_1 k_1) \right] + \frac{1}{2!}\left[f_{tt} \cdot (\alpha_1 h)^2 + 2f_{ty} \cdot (\alpha_1 h)(\beta_1 k_1) + f_{tt} \cdot (\beta_1 k_1)^2 \right] + \cdots \right\}$$

$$k_3 = hf(t_i + \underbrace{\alpha_2 h}_{\Delta t}, \ w_i + \underbrace{\gamma_1 k_1 + \gamma_2 k_2}_{\Delta y})$$

$$= h\Big\{ f(t_i, w_i) + \left[f_t \cdot (\alpha_2 h) + f_y \cdot (\gamma_1 k_1 + \gamma_2 k_2) \right]$$

$$+ \frac{1}{2!}\left[f_{tt} \cdot (\alpha_2 h)^2 + 2f_{ty} \cdot (\alpha_2 h) \cdot (\gamma_1 k_1 + \gamma_2 k_2) + f_{yy} \cdot (\gamma_1 k_1 + \gamma_2 k_2)^2 \right] + \frac{1}{3!}[\cdots] + \cdots \Big\}$$

Derivation of the RK4 solution requires both $\phi^{(4)}{}_{Taylor}$ derived from a 4th order Taylor expansion for $y(t_{i+1})$ to and an appropriate Sampled Slope expansion $\phi^{(4)}{}_{Slope}$ to match terms against. The derivation follows that for RK2, and again, by trial and error, we arrive at the "nested" form for $\phi^{(4)}{}_{Slope}$ given by the first equation on the slide. This equation contains a total of 12 parameters consisting of (i) a set of four weighting coefficients $\{a_1, a_2, a_3, a_4\}$, (ii) three parameters determining the "Δt"s $\{\alpha_1, \alpha_2, \alpha_3\}$, and (iii) a five parameter ("nesting" set) determining the "Δw"s $\{\beta_1, \gamma_1, \gamma_2, \delta_1, \delta_2, \delta_3\}$. The nested nature of this form is further illuminated by noting that each successive term contains the slope(s) of the previous term(s), as defined by the braces under the four terms of the 1st equation, *i.e.*,

$k_1 = h\, f(t_i, w_i),$

$k_2 = h\, f(t_i + \Delta t_1, w_i + \Delta w_1)$ with $\Delta t_1 = \alpha_1 h$; $\Delta w_1 = \beta_1 k_1$

$k_3 = h\, f(t_i + \Delta t_2, w_i + \Delta w_2)$ with $\Delta t_2 = \alpha_2 h$; $\Delta w_2 = \gamma_1 k_1 + \gamma_2 k_2$

$k_4 = h\, f(t_i + \Delta t_3, w_i + \Delta w_3)$ with $\Delta t_3 = \alpha_3 h$; $\Delta w_3 = \delta_1 k_1 + \delta_2 k_2 + \delta_3 k_3$

Each of these terms must now be expanded to h^4; k_1 needs no expansion; the expansion for k_2 is multiplied by h^1, so the terms inside the braces must include terms up to h^3 and we need to add the term

$$(1/3!)[\, f_{ttt}(\Delta t_2)^3 + 3\, f_{tt}\, f_y(\Delta t_2)^2 (\Delta w_2)^1 + 3\, f_t\, f_{yy}(\Delta t_2)^1 (\Delta w_2)^2 + f_{yyy}(\Delta w_2)^3].$$

Note another complication is that "Δw_2" has both k_1 and k_2 in it; k_2 in turn contains "Δw_1" which then needs to be expanded as a 2d Taylor up to to h^2. We do not intend to complete the RK4 calculation, but only wish to point out that this method is probably the most-used solution technique in existence and it is humbling to realize that its derivation is anything but trivial. Hopefully when we use this technique we will take pause for a moment and show some appreciation for the hard work that went into its derivation. This is especially significant since it was derived at a time when there were no symbolic computational tools such as Mathematica® or Maple® around to lighten the load. Moreover, this is only half of the derivation, because on the next slide we need take the 4th order Taylor expansion for $y(t_{i+1})$ and make a similar 2d Taylor expansion of the slope function $f(t,w)$ making sure to keep all terms up to and including h^4 in order to match terms.

Self-Starting Single Step Methods

2.5.1 RK4 Derivation – Matched Expansions

RK4 Derivation – Matched Expansions

1^d Taylor Expansion of y(t): $\phi_{Taylor}^{(4)} = y_i' + \dfrac{h}{2!} y_i'' + \dfrac{h^2}{3!} y_i''' + \dfrac{h^3}{4!} y_i^{iv}$

$y_i' = f$

$y_i'' = \dfrac{\partial}{\partial t}(f) + f \dfrac{\partial}{\partial y}(f) = f_t + f \cdot f_y$

$y_i''' = \dfrac{d}{dt} y_i'' \equiv \dfrac{\partial}{\partial t}(f_t + f \cdot f_y) + f \dfrac{\partial}{\partial y}(f_t + f \cdot f_y) = (f_{tt} + f_t f_y + f f_{yt}) + f(f_{ty} + f_y^2 + f f_{yy})$

$\quad = f_{tt} + f_t f_y + 2 f f_{yt} + f f_y^2 + f^2 f_{yy}$

$y_i^{iv} = \dfrac{d}{dt} y_i''' \equiv \dfrac{\partial}{\partial t}\left(f_{tt} + f_t f_y + 2 f f_{yt} + f f_y^2 + f^2 f_{yy}\right) + f \dfrac{\partial}{\partial y}\left(f_{tt} + f_t f_y + 2 f f_{yt} + f f_y^2 + f^2 f_{yy}\right)$

$\quad = (f_{ttt} + f_{tt} f_y + 3 f_t f_{yt} + f_t f_y^2) + f(5 f_y f_{yt} + 3 f_t f_{yy} + 3 f_{ytt} + f_y^3) + f^2(3 f_{yyt} + 4 f_y f_{yy}) + f^3 f_{yyy}$

Match all terms up to order h^3 (1^d Taylor y(t) & 2^d Slope) yields relations among parameters: $\alpha, \beta, \gamma, \delta$. Particular one is known as RK4

$$w_{i+1} = w_i + \dfrac{1}{6}\left[k_1 + 2(k_2 + k_3) + k_4\right]$$

$k_1 \equiv h f(t_i, w_i)$
$k_2 \equiv h f(t_i + \dfrac{h}{2}, w_i + \dfrac{1}{2} k_1)$
$k_3 \equiv h f(t_i + \dfrac{h}{2}, w_i + \dfrac{1}{2} k_2)$
$k_4 \equiv h f(t_i + h, w_i + k_3)$

Note1: k_1 "feeds" k_2 ; k_2 "feeds" k_3 ; k_3 "feeds" k_4

Note2: for $f = f(t)$ equivalent to Simpson integral !!

The 1^d Taylor expansion of $y(t_{i+1})$ yields an equation for $\phi^{(4)}_{Taylor}(t_i)$ that contains the 1st through 4th derivatives of y(t) with respect to time; however, the function y(t) is not explicitly known and needs the solution of DE y'= f(t,y) in order to express it as a function of the variable "t". As we have discussed the total derivative of f(t,y), which contains two variables, may be expressed as a derivative operator using partials in each variable and the DE slope function f to replace y'(t) wherever it occurs; we have
$$d/dt[f(t,y)] = \{\partial/\partial t + f(t,y)\partial/\partial y\} [f(t,y)] = f_t + f f_y$$
where the partials are taken as subscripts $f_t = \partial/\partial t\, [f(t,y)]$ and $f_y = \partial/\partial y\, [f(t,y)]$. The higher order derivatives are a result of expanding the operator expressions $\{\partial/\partial t + f(t,y)\partial/\partial y\}^n$ and operating on [f(t,y)]. We have computed these derivatives and listed them in the slide. When substituted back into $\phi^{(4)}_{Taylor}$ it becomes a function of both t_i and w_i with terms involving products of f, f_t, f_y, ...f_{yyy}, which can now be compared term-by-term with a fully developed expression for the sampled slope expansion $\phi^{(4)}_{Slope}$.

The result of this long and tedious process is a solution with several free parameters; one particular choice is the one known as "The RK4 Method" which is expressed by the two boxed equations at the bottom of the slide. Note that the 1st box is the "update" recursion equation which gives w_{i+1} as w_i plus a term that only depends upon the "k"s and hence only upon t_i and w_i. This added term is a simple linear combination of the 4 "k"s which are in turn computed according to the nested sequence of equations in the second box.

Note that the (red) arrows show that the terms must be calculated in the order shown because "k_1 feeds k_2 feeds k_3 feeds k_4". Also note that the stepsize is conveniently absorbed into each k-value; this is dimensionally correct since the slope function f(t,y) being a derivative has the dimension dim[y]/(length) and hence must be multiplied by a length "h" to create a quantity with the dimensions of the variable "y".

Self-Starting Single Step Methods

2.5.2 Runge-Kutta Summary & Simpson

Runge-Kutta Summary & Simpson

- IC generates unique solution: $w_0 = \alpha$
  only need previous value
  w_i to find w_{i+1}

  $w_1 = w_0 + h\varphi(w_0)$
  $w_2 = w_1 + h\varphi(w_1)$, etc.

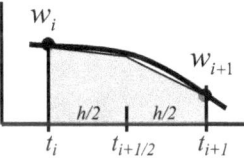

- RK4 integrates y' with cubic exactness
- Requires 4 'sample' slopes $k_i = h\, m_i$
- No derivatives are required

$$w_{i+1} = w_i + \frac{1}{6}\left[k_1 + 2(k_2 + k_3) + k_4\right]$$

$$k_1 \equiv hf(t_i, w_i)\ ;\ k_2 \equiv hf(t_{i+1/2}, w_i + \frac{1}{2}k_1)$$

$$k_3 \equiv hf(t_{i+1/2}, w_i + \frac{1}{2}k_2)\ ;\ k_4 \equiv hf(t_i + h, w_i + k_3)$$

- If $f(t,y(t)) = g(t)$: fcn of t only
  → Simpson Integration Rule

$$y = \int_{t_0}^{t_0+h} g(t)dt = \frac{h/2}{3}\left\{\underbrace{g(t_0)}_{=m_1} + 2\left[\underbrace{g(t_0+h/2)}_{=m_2} + \underbrace{g(t_0+h/2)}_{=m_3}\right] + \underbrace{g(t_0+h)}_{=m_4}\right\}$$

$$\xrightarrow[h\to 2h]{} \frac{h}{3}\{g(t_0) + 4g(t_0+h) + g(t_0+2h)\}$$

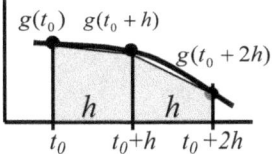

The RK4 method is known as a self-starting one because the solution only requires a single start value, the IC $w_0 = \alpha$ of the IVP. A typical integration panel is shown with the three time points $\{t_i, t_{i+½}, t_{i+1}\}$ at which the slope function is to be evaluated; care must be taken to use the correct y-values for each time point as follows:

k_1 is evaluated at the first time point t_i and uses a y-value of w_i
k_2 and k_3 are both evaluated at $t_{i+½}$ but because of "nesting" they use quite different y-values;
k_2 uses a y-value $w_i + ½\, k_1$, while k_3 uses a y-value $w_i + ½\, k_2$
k_4 is evaluated at the third time point t_{i+1} and uses a y-value of $w_i + k_3$

Thus although there are only three time points used the slope function is evaluated 4 times, twice at the middle time point.

The RK4 method integrates the DE for y(t) with cubic exactness in "t", which means explicitly that if $f(t,y) = t^3$, the solution numerical solution has no truncation error. In fact, in the general case where $f(t,y) = g(t)$ is only a function of "t" the problem reduces to a Simpson quadrature integral; for i=0 RK4 gives

$$w_1 - w_0 = (h/6)\{g(t_0) + 2[g(t_0+h/2) + g(t_0+h/2)] + g(t_0+h)\}$$

Upon letting h=> 2h corresponding to the usual h stepsize for a single Simpson panel we have

$$w_1 - w_0 = (2h)/6\ \{g(t_0) + 2[g(t_0+2h/2) + g(t_0+2h/2)] + g(t_0+2h)\}$$
$$= (h/3)\{g(t_0) + 4g(t_0+h) + g(t_0+2h)\}$$

which is the Simpson 1-panel result.

Self-Starting Single Step Methods

2.5.3 RK4 Example – 2 Steps

RK4 Example – 2 Steps

$$y' = -ty^2 \;;\; y(2) = 1 \;;\; \text{Exact}: y(t) = \frac{2}{t^2 - 2}$$

$w_0 \to w_1$

$h = 0.1 \;;\; t_0 = 2 \;;\; w_0 = y_0 = 1$

$k_1 \equiv hf(t_0, w_0) = 0.1(-2 \cdot 1^2) = -.2$

$k_2 \equiv hf(t_0 + \frac{h}{2}, w_0 + \frac{1}{2}k_1) = 0.1\left[-(2 + \frac{.1}{2}) \cdot (1 + \frac{1}{2}(-.2))^2\right] = -0.16605$

(with $t_0 + h/2$ and $w_0 + k_1/2$ indicated)

$k_3 \equiv hf(t_0 + \frac{h}{2}, w_0 + \frac{1}{2}k_2) = 0.1\left[-(2 + \frac{.1}{2}) \cdot (1 + \frac{1}{2}(-0.16605))^2\right] = -.172373$

$k_4 \equiv hf(t_0 + h, w_0 + k_3) = 0.1\left[-(2 + .1) \cdot (1 + (-.172373))^2\right] = -0.143843$

$w_1 = w_0 + \frac{1}{6}[k_1 + 2(k_2 + k_3) + k_4] = 1 + [-.2 + 2(-0.16605 - .172373) - 0.143843]/6 = \boxed{0.829885}$

$y_{exact}(2.1) = \frac{2}{2.1^2 - 2} = \boxed{0.829875}$ ←——— Compare: 4 sd

$w_1 \to w_2$

$h = 0.1 \;;\; t_1 = 2.1 \;;\; w_1 = 0.829885$

$k_1 \equiv hf(t_1, w_1) = 0.1(-(2.1) \cdot (0.829885)^2) = -0.1446289$

$k_2 \equiv hf(t_1 + \frac{h}{2}, w_1 + \frac{1}{2}k_1) = 0.1\left[-(2.1 + \frac{.1}{2}) \cdot (0.829885 + \frac{1}{2}(-0.1446289))^2\right] = -0.1233913$

$k_3 \equiv hf(t_1 + \frac{h}{2}, w_1 + \frac{1}{2}k_2) = 0.1\left[-(2.1 + \frac{.1}{2}) \cdot (0.829885 + \frac{1}{2}(-0.1233913))^2\right] = -0.1268747$

$k_4 \equiv hf(t_1 + h, w_1 + k_3)k_1 = 0.1\left[-(2.1 + .1) \cdot (0.829885 + (-0.1268747))^2\right] = -0.10871989$

$w_2 = w1 + \frac{1}{6}[k_1 + 2(k_2 + k_3) + k_4] = 0.829885 + [-0.1446289 + 2(-0.1233913 - 0.1268747) - 0.10871989]/6 = \boxed{0.7042382}$

$y_{exact}(2.2) = \frac{2}{2.2^2 - 2} = \boxed{0.70422535}$ ←——— Compare: 4 sd

Here is a detailed calculation of two steps of the RK4 solution for the IVP $\{y' = -ty^2 \;;\; y(2) = 1\}$. From the IC, we have $t_0 = 2$, $w_0 = y_0 = 1$, and taking a stepsize $h = 0.1$, we first compute the "k"s for the step from $w_0 \to w_1$ taking care to "feed the value of k_1 into the calculation for k_2" *etc.* down the line and finally apply the update formula to find $w_1 = .829885$. This value is good to 4 sd compared with the exact solution; the next step from $w_1 \to w_2$ proceeds in the same manner yielding $w_2 = .7042382$ again good to 4 sd.

Clearly, automation of these computations with a calculator or computer program is called for, but it is good to "get close to the ground" at least once in order to understand the pitfalls which can and will occur when you go to program this algorithm or another on your own. For example, it is easy to confuse the labels and evaluate the slope function at the wrong (t_i, w_i) coordinates or to make an error nesting the "k"s in the multiple-indexed environment of a computer program. Making a simple "run through" with the help of a handheld calculator can catch such errors because you have to patiently go through each and every step and in so doing, the nature of your indexing or other error may pop out like a jack in the box (or not). In any case, it builds character and you will have a greater appreciation for the work that others have done.

Self-Starting Single Step Methods

2.6 Computational Effort Trade-off: Order vs. Stepsize

Computational Effort Trade-off: Order *vs.* Stepsize

- Output at: t, t+H, t+2H, ...
- Stepsize $h = H/r$ vs. Order of Method
- Computational Effort = total # fcnal evals N_F
- High Order fewer steps r & More evals/step N_{Step}
- Low Order More steps r & fewer evals/step N_{Step}

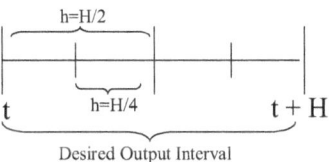

$$N_F = N_{Step} \cdot r$$

Functional Evaluations per Step

h	Euler O(h)	RK2 O(h²)	RK4 O(h⁴)
H	1	2	④
H/2	2	④	8
H/4	④	8	16

Method Order →

Compare
RK4 1 H-Step
RK2 2 H/2-Steps
Euler 4 H/4-Steps

RK7 requires =9,10 eval/step
RK6 requires =7,8 eval/step
RK5 requires 6 eval/step
RK4 →Diminishing returns

Method Order: n	h¹	h²	h³	h⁴	h⁵	h⁶	h⁷
Fcn Evals: N_F	1	2	3	4	6	7,8	9,10

In general when we solve an IVP, we desire output at specified time intervals, t, t+ H, t+2H,... and we would like to minimize the computational effort in obtaining these outputs. Because functional evaluations are most costly, the total computational cost of an algorithm can be approximated by N_F the total number of slope function $f(t_i,w_i)$ evaluations. If there are r steps, each requiring N_{Step} functional evaluations, then the total number of evaluations is $N_F = r\, N_{Step}$ and this corresponds to fact that lower order methods require r steps of stepsize H/r to attain the same truncation error as a higher order method (See slide#2-10).

The 1st table lists the total number of functional evaluations N_F for the Euler, RK2, and RK4 methods across the columns for three different stepsizes, H, H/2, and H/4 in the rows. The circled numbers in the table all require the same computational effort N_F =4 which occurs for r= 4 Euler steps of H/4, 2 RK2 steps of H/2 and 1 RK4 step of H.

Note we previously found that in terms of truncation error the RK4 method is most efficient of the three methods (See slide#2-10) since approximately 3 "H/3 steps" provides a 100-fold improvement in truncation error, whereas Euler and RK2 require 100 and 10 steps respectively. The question is "can we gain further advantage by going to higher order RK5, RK6, ... methods?"

The table, almost certainly computed with great difficulty, gives the answer to this question by showing the linear progression of functional evaluations N_F with increased truncation error order stops at RK4; RK5 requires N_F =6, RK6 requires N_F =7 or 8, and RK7 requires N_F =9 or 10 (depending upon family member.) The "diminishing returns" of methods beyond RK4 is the reason for its popularity, *i.e.*, it is both relatively simple and is at the "cusp" of computational efficiency.

2.6.1 Computational Effort Trade-off: Example

Computational Effort Trade-off: Example

$$y' = y - t^2 + 1 \;;\; t \in [0,.2] \;;\; y(0) = 0.5$$

- Euler, RK2, RK4 have same Computational Effort: $N_F = 4$

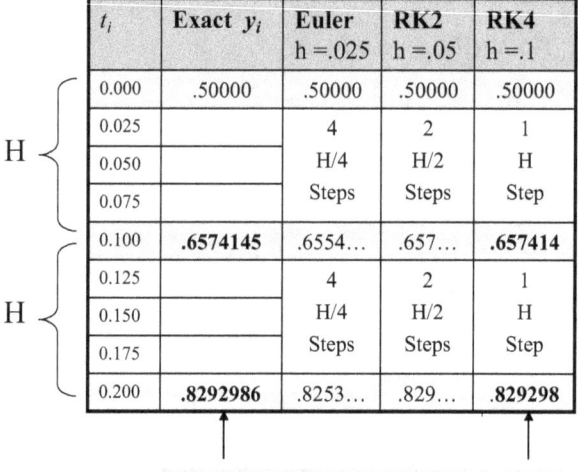

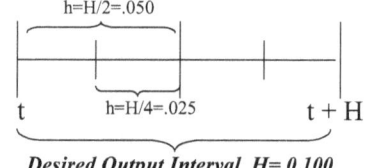

Desired Output Interval H = 0.100

Compare Methods: RK4 has ~6 sd

In the last slide we established that RK4 is the most computationally efficient of the Runge-Kutta methods and here we give an explicit comparison of the output at specified time intervals, t, t+ H, t+2H, for Euler, RK2, and RK4. Results for the IVP is {y'= y - t² + 1 ; t∈[0, 0.2] ; y(0) = 0.5} are shown in the short table for two H=0.1 outputs. The number and size of steps are indicated in each column (but not shown) and are chosen to make the total number of functional evaluations between outputs the same for each method.

Comparing the outputs for the three methods with the exact values in col#2 shows that, as expected, given the same computational burden, the RK4 method is superior to the other two attaining nearly 5 sd accuracy compared with 2 or 3 for the other two. The next slide gives a detailed comparison of the Euler and RK4 methods showing the steps in between for Euler and the increased accuracy for RK4 when they both have the same stepsize h.

2.6.2 Detailed Comparison of Euler and RK4

Detailed Comparison of Euler *and* RK4 Methods

IVP: $y' = y - t^2 + 1$; $t \in [0,1]$; $y(0) = 0.5$

$y_{exact} = (t+1)^2 - 0.5*\exp(t)$

			Euler h=.05		RK4 h=.05		RK4 h=.25	
i	ti	y_exact	wi	|yi-wi|	wi	|yi-wi|	wi	|yi-wi|
0	0	0.5	0.5	0	0.5	0	0.5	0
1	0.05	0.57686	0.575	0.0018645	0.57686	5.20E-09		
2	0.1	0.65741	0.65362	0.0037895	0.65741	1.06E-08		
3	0.15	0.74158	0.73581	0.0057766	0.74158	1.62E-08		
4	0.2	0.8293	0.82147	0.0078271	0.8293	2.20E-08		
5	0.25	0.92049	0.91055	0.0099422	0.92049	2.80E-08	0.92047	1.61E-05
6	0.3	1.0151	1.0029	0.012123	1.0151	3.43E-08		
7	0.35	1.113	1.0986	0.014371	1.113	4.08E-08		
8	0.4	1.2141	1.1974	0.016688	1.2141	4.75E-08		
9	0.45	1.3183	1.2993	0.019074	1.3183	5.45E-08		
10	0.5	1.4256	1.4041	0.021531	1.4256	6.18E-08	1.4256	3.56E-05
11	0.55	1.5359	1.5118	0.02406	1.5359	6.93E-08		
12	0.6	1.6489	1.6223	0.026662	1.6489	7.71E-08		
13	0.65	1.7647	1.7354	0.029337	1.7647	8.52E-08		
14	0.7	1.8831	1.851	0.032086	1.8831	9.35E-08		
15	0.75	2.004	1.9691	0.03491	2.004	1.02E-07	2.0039	5.90E-05
16	0.8	2.1272	2.0894	0.037811	2.1272	1.11E-07		
17	0.85	2.2527	2.2119	0.040787	2.2527	1.20E-07		
18	0.9	2.3802	2.3364	0.043839	2.3802	1.30E-07		
19	0.95	2.5096	2.4627	0.046968	2.5096	1.40E-07		
20	1	2.6409	2.5907	0.050173	2.6409	1.50E-07	2.6408	8.71E-05

Here is an output table for the same ODE as the last slide but for a larger domain, *viz.*, [0, 1]. The table compares the Euler and RK4 methods first using a stepsize h = 0.05 and then increasing the stepsize for RK4 by a factor of five to h = 0.25. (The MatLab® scripts with script generated outputs are given for Euler on Slides# 8-2, 8-3 and for RK4 on Slides# 8-4, 8-5.) The iteration index is shown for convenience in column #1, and the time t_i and exact solution y_{exact} are displayed respectively in columns #2 and #3. The remaining columns give Euler and its error in #4, 5 (shaded), Rk4 and its error for stepsize .05 in #6, 7 and for large stepsize .25 in #7, 8 (shaded). Note that Euler error increases from 1.8×10^{-3} in the first step to .05 in step #20, while the RK4 is very accurate having an error that ranges from 10^{-9} to 10^{-7}; moreover, even when the RK4 stepsize is increased five-fold to .25 the error is always of order 10^{-5} which is orders of magnitude more accurate than the Euler method. Thus, it is clear that the RK4 method is superior to the Euler method and is to be preferred. The Euler method is still important because it is easily derived and understood and it is also useful for theoretical arguments and proofs, but is not an efficient or effective method for numerical solutions to ODEs.

2.7 Error Control – Variable Stepsize

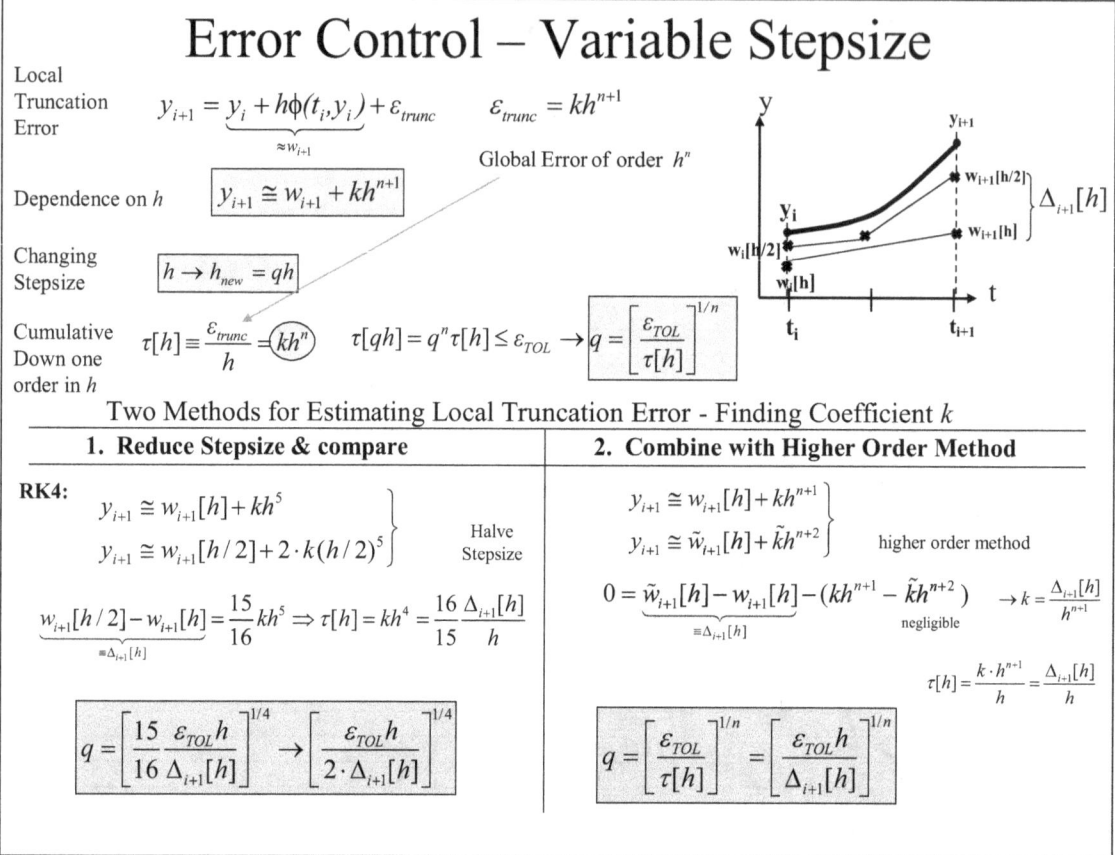

Using a fixed stepsize is wasteful, especially when there are large variations in f(t,y); it is computationally more efficient to increase or decrease the stepsize as needed to attain a fixed error tolerance just as was done for numerical quadratures.

In general, the self-starting method of solving an IVP takes a single start value w_0 at t=a, and sequentially updates the y-value *via* a recursion update equation: $w_{i+1} = w_i + h\, \phi(t_i, w_i)$ for i=0,1,2, ... The function $\phi(t_i, w_i)$ is a linear combination of sampled slopes m_i which defines the "method" as Euler, RK2, RK4, *etc.* . If we replace w_i by y_i in the "ϕ-method" and use $\phi(t_i, y_i)$ to estimate the formal integral for $y_{i+1} - y_i$ we obtain

$$y_{i+1} = y_i + h\, \phi(t_i, y_i) + \varepsilon_{trunc}$$

where the truncation error is $\varepsilon_{trunc} = kh^{n+1}$.

The local truncation error at t_{i+1} is defined as

$$\varepsilon_{i+1} = y_{i+1} - w_{i+1} \approx y_{i+1} - [\, y_i + h\, \phi(t_i, y_i)\,] = \varepsilon_{trunc} = kh^{n+1}$$

The quantity $\tau[h] = \varepsilon_{trunc}[h]/h = kh^n$ is a measure of the cumulative truncation error; if we change the stepsize from h to q*h and require the resulting cumulative error to be less than some fixed value ε_{Tol} then

$$\tau[qh] = k(qh)^n = q^n k h^n = q^n \tau[h] < \varepsilon_{Tol}$$

which is solved for the stepsize ratio $q = \{\varepsilon_{Tol}/\tau[h]\}^{1/n}$

The two methods of computing the stepsize needed to maintain a fixed value ε_{Tol} are to compute results again with either (i) a reduced stepsize or (ii) a higher order method, as shown in the lower panel of the slide. On the LHS we halve the stepsize for RK4 and subtract the two computed values to estimate $\tau[h] = k\, h^4$ which upon substitution into the stepsize ratio equation yields the boxed equation for q. On the RHS we subtract results from two methods of different order and drop the truncation error of the higher order term which is negligible compared with that of the lower order term; this yields $\tau[h] = kh^{n+1} = \Delta_{i+1}[h]/h$ and again substitution into the stepsize ratio equation yields the boxed equation for q on the right.

2.7.1 RK4 & RKF45 Comparison

RK4 & RKF45 Comparison

- **RK4 Error Control Requires 11 fcnal evals**
  - 4 $f(t,y)$ evals for 1 H-step &
  - 8 $f(t,y)$ evals for 2 H/2-steps
  - Only one eval is common $f(t_i, w_i)$

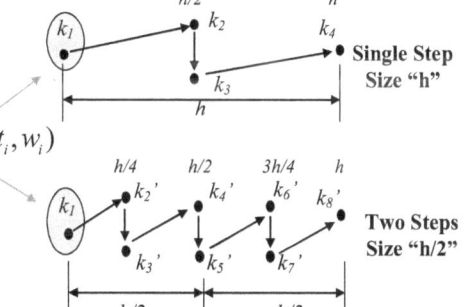

  - → 12−1=11 evals

$q = \left[\dfrac{\varepsilon_{TOL} h}{2\Delta_{i+1}[h]} \right]^{1/4}$ Stepsize Ratio $\quad \Delta_{i+1}[h] = |w_{i+1}[h] - w_{i+1}[h/2]|$

- **RKF45 Error Control Requires 6 fcnal evals**
  - Combines 4th and 5th Order with common fcnal evals

$y_{i+1} \cong w_{i+1}[h] + kh^5$ 4th Order Global Error

$y_{i+1} \cong \widetilde{w}_{i+1}[h] + kh^6$ 5th Order Global Error

$q = \left[\dfrac{\varepsilon_{TOL} h}{\widetilde{\Delta}_{i+1}[h]} \right]^{1/5}$ Stepsize Ratio

$\widetilde{\Delta}_{i+1}[h] = |w_{i+1}[h] - \widetilde{w}_{i+1}[h]|$

In this slide we compare computational efficiency of the the **halving method** applied to RK4 with the **higher order method** used for the RKF45. The latter is a variation of the RK4 method developed by Fehlberg that uses a related pair of 4th and 5th order methods as detailed on the next slide.

The RK4 method is used twice: first with a single step of size h requiring 4 evaluations of f(t,y) and then with 2 steps of size h/2 requiring 2(4)=8 evaluations of f(t,y). Only the first evaluation is common between these two calculations as illustrated in the figure, and this leaves a total of 12-1 = 11 distinct f(t,y) evaluations.

On the other hand the RKF45 method given on the next slide only has a total 6 distinct functional evaluations in the 4th and 5th order computations and hence is computationally more efficient than the RK4 method with halving.

Self-Starting Single Step Methods

2.7.2 Runge-Kutta-Fehlberg Method

Runge-Kutta-Fehlberg Method

Step Eqns:

$$\widetilde{w}_{i+1} = w_i + \frac{16}{135}k_1 + \frac{6656}{12825}k_3 + \frac{28561}{56430}k_4 - \frac{9}{50}k_5 + \frac{2}{55}k_6 \qquad O(h^5)$$

$$w_{i+1} = w_i + \frac{25}{216}k_1 + \frac{1408}{2565}k_3 + \frac{2197}{4104}k_4 - \frac{1}{5}k_5 \qquad O(h^4)$$

Note: k_2 is missing

Step Coefficients:

$$k_1 = hf(t_i, w_i)$$

$$k_2 = hf(t_i + \frac{1}{4}h, w_i + \frac{1}{4}k_1)$$

$$k_3 = hf(t_i + \frac{3}{8}h, w_i + \frac{3}{32}k_1 + \frac{9}{32}k_2)$$

$$k_4 = hf(t_i + \frac{12}{13}h, w_i + \frac{1932}{2197}k_1 - \frac{7200}{2197}\mathbf{k_2} + \frac{7296}{2197}k_3)$$

$$k_5 = hf(t_i + h, w_i + \frac{439}{216}k_1 - 8\mathbf{k_2} + \frac{3680}{513}k_3 - \frac{845}{4104}k_4)$$

$$k_6 = hf(t_i + \frac{1}{2}h, w_i - \frac{8}{27}k_1 + 2\mathbf{k_2} - \frac{3544}{2565}k_3 + \frac{1859}{4104}k_4 - \frac{11}{40}k_5)$$

Only 6 fcnal evals $f(t,y)$ for $k_1, \ldots, k_6$

k_2 occurs in "nesting"

The Runge-Kutta-Fehlberg method computes $\widetilde{w}_{i+1}$ using the 5th order formula and the set of six "k"s $\{k_1, k_2, k_3, k_4, k_5, k_6,\}$ given below and also computes w_{i+1} using the 4th order formula and the same six "k"s. The difference $\Delta_{i+1}[h] = w_{i+1} - \widetilde{w}_{i+1}$ is computed and the **order n=4 corresponding to the lower order computation** is used to compute the stepsize ratio according to

$$q = (\varepsilon_{Tol} h / \Delta_{i+1}[h])^{1/4}$$

Note that k_2 does not occur in either of the update equations, but it does occur in the nested computation of $k_3, k_4, k_5,$ and k_6. Also note since there is some cost in changing the stepsize ratio practical limits (gained by experience) are usually placed on how large or how small the ratio should be so the algorithm is not unnecessarily changing the stepsize for every small variation.

2.7.3 Example of RKF45 with Stepsize Control

Example of RKF45 with Stepsize Control

IVP: $y' = y - t^2 + 1$; $t \in [0,4]$; $y(0) = 0.5$ *Compare with Slide#2-24*

$y_{exact} = (t+1)^2 - 0.5 \cdot \exp(t)$

index	ti	wi	h	$\tau[h]$	y_exact	|yi-wi|
0	0	0.5	0.25	NaN	0.5	0
1	0.25	0.92049	0.25	6.21E-06	0.92049	1.31E-06
2	0.48655	1.3965	0.23655	4.49E-06	1.3965	2.57E-06
3	0.72933	1.9537	0.24278	4.27E-06	1.9537	4.17E-06
4	0.97933	2.5864	0.25	3.77E-06	2.5864	6.19E-06
5	1.2293	3.2605	0.25	2.44E-06	3.2605	8.50E-06
6	1.4793	3.9521	0.25	7.22E-07	3.9521	1.11E-05
7	1.7293	4.6308	0.25	1.48E-06	4.6308	1.41E-05
8	1.9793	5.2575	0.25	4.31E-06	5.2575	1.73E-05
9	2.2293	5.7818	0.25	7.94E-06	5.7818	2.07E-05
10	2.4518	6.1103	0.22244	7.99E-06	6.1103	2.45E-05
11	2.6494	6.2454	0.19762	7.12E-06	6.2453	2.87E-05
12	2.8301	6.1961	0.18068	6.63E-06	6.1961	3.33E-05
13	2.9983	5.9608	0.16821	6.35E-06	5.9607	3.85E-05
14	3.1566	5.5322	0.15828	6.17E-06	5.5322	4.43E-05
15	3.3066	4.901	0.15	6.05E-06	4.901	5.06E-05
16	3.4494	4.0563	0.14287	5.95E-06	4.0562	5.76E-05
17	3.5861	2.9863	0.13663	5.88E-06	2.9862	6.54E-05
18	3.7171	1.6784	0.13106	5.82E-06	1.6784	7.38E-05
19	3.8432	0.11973	0.12606	5.77E-06	0.11964	8.31E-05
20	3.9647	-1.7035	0.1215	5.72E-06	-1.7036	9.32E-05
21	4	-2.299	0.035323	4.82E-08	-2.2991	9.66E-05

Variable Stepsize Control Parameters

$$h^{new} = q \cdot h^{old}$$

$$q = .84 \cdot \left[\frac{\varepsilon_{TOL}}{\tau[h]} \right]^{1/4}$$

$$\tau[h] = |\Delta w_{i+1}| / h$$
$$= |w_{i+1} - \widetilde{w}_{i+1}| / h$$

$\varepsilon_{TOL} = 10^{-5}$

Stepsize Limits

$h_{min} = .02$
$h_{max} = .25$

Here is an output table for the same ODE as on Slide#2-24 but using the Runge-Kutta-Felhberg (RKF45) method with stepsize control. As a practical matter, the stepsize is constrained to be within a certain range of values and an error tolerance TOL is specified; in this case the variable stepsize is constrained between $h_{max} = .25$ and $h_{min} = .02$ and $\varepsilon_{TOL} = 10^{-5}$. Again, the iteration index is shown for convenience in column #1, and the time t_i in column #2; column #3 shows the numerical estimate w_i, #4 the stepsize h, and #5 the global truncation error $\tau[h]$; column #6 displays the exact solution y_{exact} and #7 the absolute error $|y_{exact}(t_i) - w_i|$. (The MatLab® script for RKF45 is given on Slides# 8-6, 8-7.)

The quantity $\tau[h] = |\Delta w_{i+1}|/h = |w_{i+1} - \widetilde{w}_{i+1}|/h$ is the global or accumulated truncation error obtained from the 4th and 5th order solutions in the RKF45 formulation. It is used together with the specified error tolerance ε_{TOL} to compute the stepsize ratio $q = 0.84 \cdot (\varepsilon_{TOL}/\tau(h))^{1/4}$, thereby effecting stepsize control *via* "Combine with Higher Order Method" (Method 2 of Slide#2-25). This method is uniquely suited to the RKF45 formalism because of the common slope terms in the two different orders; this requires only 6 functional evaluations for both orders. On the other hand, the RK4 method uses "Reduce Stepsize and Compare Method" (Method 1 of Slide#2-25) by halving stepsize and requires 11 functional evaluations which is far less efficient than RKF45.

The first stepsize change occurs from index 1 to index 2 and is computed as follows $q = 0.84 \cdot (\varepsilon_{TOL}/\tau(h))^{1/4} = 0.84 \cdot (10^{-5}/6.21 \cdot 10^{-6})^{1/4} = .946251$; hence $h^{new} = 0.25 * .946251 = .23656 = h_2$. Inspection of the table shows that for early times the maximum stepsize h=0.25 can be used for the most part, but in order to maintain the error tolerance as the time increases (near bottom of the table), the stepsize must be reduced to .035323 at $t_i = 4$ which is close to the minimum value $h_{min} = .02$. Note also that since the table output values are not at uniform time intervals, the process of fitting an interpolation polynomial to it is more computationally intensive.

3 Multistep Methods

Multi-Step Methods

3.1 Introduction to Multistep Methods

Introduction to Multistep Methods

- **4th Order Adams–Bashforth - Explicit 4-Step**
  - *4 Start Values* (RK4): $w_0 = \alpha$; $w_1 = \alpha_1$; $w_2 = \alpha_2$; $w_3 = \alpha_3$

  $$w_{i+1} = w_i + \frac{h}{24}\left[55 f_i - 59 f_{i-1} + 37 f_{i-2} - 9 f_{i-3}\right]$$

  $$f_i = f(t_i, w_i) \; ; \; i = 3, 4, \cdots, n-1$$

  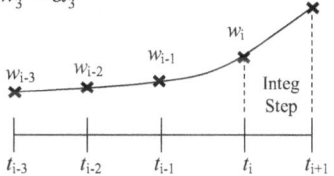

- **3rd Order Adams–Moulton - Implicit 3-Step**
  - *3 Start Values* (RK4): $w_0 = \alpha$; $w_1 = \alpha_1$; $w_2 = \alpha_2$

  $$w_{i+1} = w_i + \frac{h}{24}\left[9 f_{i+1} + 19 f_i - 5 f_{i-1} + f_{i-2}\right]$$

  $$f_i = f(t_i, w_i) \; ; \; i = 2, 3, \cdots, n-1$$

  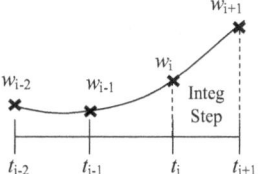

Implicit Example:

$$f_{i+1} = f(t_{i+1}, w_{i+1}) = -t_{i+1} \cdot w_{i+1}^2$$

$$w_{i+1} = w_i + \frac{h}{24}\left(-9 t_{i+1} \cdot w_{i+1}^2\right) + \frac{h}{24}\left(19 f_i - 5 f_{i-1} + f_{i-2}\right)$$

*Variable w_{i+1} occurs on both sides and **is not linear** so need Newton-Raphson to solve quadratic for w_{i+1}*

Multistep methods are not self-starting because the initial condition must be augmented by several other start values; in order to proceed, these additional start values must first be generated by a self-starting method such as RK4. Multistep methods fall into two categories depending upon whether the integration step from t_i to t_{i+1}

(i) requires only prior values w_{i-3}, w_{i-2}, w_{i-1}, w_i as in the upper panel (explicit), or
(ii) uses values including w_{i+1} at the desired output point (implicit) as in the lower panel.

The upper panel shows the 4th order Adams-Bashforth (AB4) method which requires three additional start values in order to complete the integration step from t_i to t_{i+1}. The sequence of outputs starts with w_4 which requires $w_0=\alpha$, w_1, w_2, and w_3 to start the method; in this regard, note that the index scheme starts with i =3 (not i=0). The formula uses the standard notation for the indexed time $t_i = t_0 + i\,h$, and $f_i = f(t_i, w_i)$, so for example, f_{i-2} means the slope function is evaluated at (t_{i-2}, w_{i-2}), which when written out explicitly yields $f_{i-2} = f(t_{i-2}, w_{i-2})$.

The lower panel shows the 3rd order Adams-Moulton (AM3) method which requires two additional start values in order to complete the integration step from t_i to t_{i+1}. The sequence of outputs starts with w_3 which requires $w_0=\alpha$, w_1, and w_2 to start the method. Since w_{i+1} occurs on both sides of the equation and $f(t_{i+1}, w_{i+1})$ is generally a non-linear function such as $f(t_{i+1}, w_{i+1}) = - t_{i+1} * w^2_{i+1}$, implicit methods require a root-finding technique such as Newton-Raphson to find the value of w_{i+1}.

3.2 Multistep Formulation – Newton Polynomials

Multistep Formulation – Newton Polynomials

- Integrate $\quad y_{i+1} = y_i + \int_{t_i}^{t_{i+1}} f(t, y(t))dt$
- Fit Newton Backward Interpolating polynomial through a set of m nodes

$$\underbrace{(t_i, y_i), (t_{i-1}, y_{i-1}), \cdots, (t_{i-(m-1)}, y_{i-(m-1)})}_{m\text{-nodal points}}$$

$$f(t, y(t)) = P_{m-1}(t) + \underbrace{\frac{f^{(m)}(\xi, y(\xi))}{m!}(t - t_i)(t - t_{i-1})\cdots(t - t_{i-(m-1)})}_{R_{(m-1)}(t, \xi(t))}$$

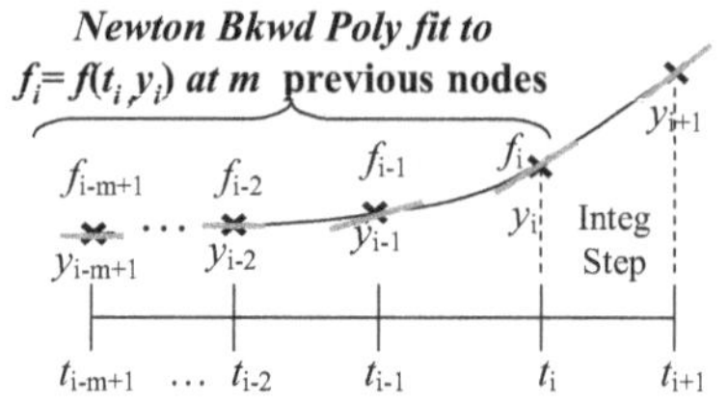

Newton Bkwd Poly fit to $f_i = f(t_i, y_i)$ at m previous nodes

- Substitute $\quad y_{i+1} = y_i + \int_{t_i}^{t_{i+1}} P_{m-1}(t)dt + \int_{t_i}^{t_{i+1}} R_{m-1}(t, \xi(t))dt$

Backward Differences
$f_i \equiv f(t_i, y_i)\,; \;\nabla f_i = f_i - f_{i-1}\,;$
$\nabla^2 f_i = \nabla(f_i - f_{i-1}) = \nabla f_i - \nabla f_{i-1}$
$= (f_i - f_{i-1}) - (f_{i-1} - f_{i-2}) = f_i - 2f_{i-1} + f_{i-2}$

- Uniform Grid $\quad t = t_i + sh\,; \; dt = hds$
- Backward Interpolation (from t_i to t_{i-m+1})

$$y_{i+1} \cong y_i + \int_{t_i}^{t_{i+1}} \underbrace{\sum_{k=0}^{m-1}(-1)^k \binom{-s}{k} \nabla^k f(t_i, y_i)}_{\text{Backward } \Delta \text{ differences abt } t_i} h \cdot ds + h^{m+1} f^{(m)}(\xi_i, y(\xi_i)) \cdot \int_0^1 (-1)^m \binom{-s}{m} ds$$

Yields

$$y_{i+1} \cong y_i + h\sum_{k=0}^{m-1} \nabla^k f(t_i, y_i) I_k + \underbrace{h^{m+1} f^{(m)}(\xi, y(\xi)) I_m}_{\text{Trunc. Error } R_{(m-1)}} \Longrightarrow$$

Integrals
$$I_k \equiv (-1)^k \int_0^1 \binom{-s}{k} ds\,;$$

Difference Equation
$$w_{i+1} = w_i + h \cdot \underbrace{\sum_{k=0}^{m-1} \nabla^k f(t_i, w_i) \cdot I_k}_{\text{Method: } \phi(t_i, w_i)}$$

The multistep solution to a DE is a result of replacing the slope function f(t,y(t)) with an m-point polynomial $P_{(m-1)}(t)$ and its associated truncation error term $R_{(m-1)}(t,\xi(t))$. As illustrated in the figure, the polynomial is fit to the "m" slope function "values" (red slopes f(t_i,y_i)) preceding the desired output point y_{i+1} and is given by a Newton Backward Interpolating polynomial of order (m-1) for the slope function f(t_i,y_i). Note previous applications of Newton interpolations were on y_i=f(t_i); however, here we have a DE y'= f(t, y(t)) so the nodal points and values are t_i and f_i = f(t_i, y(t_i)) respectively.

Recalling the Δ-formulation for uniformly spaced points with stepsize h, we transform the integrals using t = t_i + s·h and dt = hds and find the result shown in the boxed equation with the coefficients given by the integrals I_k of $^{-s}C_k$ (binomial coefficients) integrated over s from 0 to 1. The backward ∇ is over the index "i" and, for example, the first backward difference is given explicitly by ∇f_i = f_i – f_{i-1} = f(t_i,y_i) - f(t_{i-1}, y_{i-1}) and is simply the difference in the slope function evaluated at the two indexed (t,y) points. The error term requires integration over an unknown function ξ(t), so we must take some upper bound estimate of it over the interval of interest and bring it outside the integral or equivalently apply the weighted mean theorem to evaluate it at some arbitrary point which is subsequently maximized. The integrals I_k are tabulated on the next slide.

Multi-Step Methods

3.2.1 Multistep Coefficient Integrals I_k

Multistep Coefficient Integrals I_k

- **Integral** $I_k \equiv (-1)^k \int_0^1 \binom{-s}{k} ds$

- **Example** $I_3 \equiv (-1)^3 \int_0^1 \binom{-s}{3} ds = -\int_0^1 \frac{(-s)(-s-1)(-s-2)}{3!} ds = +\frac{1}{6}\int_0^1 (s^3 + 3s^2 + 2s)\,ds = \frac{3}{8}$

- **Table**

k	0	1	2	3	4	5
I_k	1	1/2	5/12	3/8	251/720	95/288

- **m-step Method**

$$y_{i+1} \cong y_i + h\left[f(t_i, y_i) + \frac{1}{2}\nabla f(t_i, y_i) + \frac{5}{12}\nabla^2 f(t_i, y_i) + \frac{3}{8}\nabla^3 f(t_i, y_i) + \frac{251}{720}\nabla^4 f(t_i, y_i) + \cdots \right] + R^{(m)}(\xi)$$

$$R^{(m)}(\xi) = h^{m+1} \cdot f^{(m)}(\xi_i, y(\xi_i)) I_m$$

- **m=3 Example** $f_i \equiv f(t_i, y_i)$; $\nabla f_i = f_i - f_{i-1}$; $\nabla^2 f_i = \nabla(f_i - f_{i-1}) = \nabla f_i - \nabla f_{i-1} = (f_i - f_{i-1}) - (f_{i-1} - f_{i-2}) = f_i - 2f_{i-1} + f_{i-2}$

$$y_{i+1} \cong y_i + h\left[f_i + \frac{1}{2}\underbrace{[f_i - f_{i-1}]}_{1^{st}\text{ Bkwd Diff}} + \frac{5}{12}\underbrace{[f_i - 2f_{i-1} + f_{i-2}]}_{2^{nd}\text{ Bkwd Diff}} \right] + h^4 f^{(3)}(\xi_i, y(\xi_i)) I_3$$

$$y_{i+1} \cong y_i + \frac{h}{12}\left[\underbrace{(12+6+5)}_{23} f_i - \underbrace{(6+10)}_{16} f_{i-1} + 5 f_{i-2} \right] + \frac{3}{8} h^4 f^{(3)}(\xi_i, y(\xi_i))$$

$w_0 = \alpha$

$w_{i+1} = w_i + \frac{h}{12}\left[23 f_i - 16 f_{i-1} + 5 f_{i-2} \right]$

$\varepsilon_{trunc} = +\frac{3}{8} h^4 f^{(3)}(\xi_i, y(\xi_i))$

$\xi_i \in [t_i, t_{i+1}]$

The required coefficient integrals I_k are tabulated for the first five values of the index k and substituted into the General Multistep Formula of the last slide to yield a more useful intermediate result for the *m-step method* given by the boxed formula in the center of the slide.

For m=3 we have computed the necessary backward differences, substituted them into the m-step formula, collected terms, and made the transition to a difference equation by letting $y_i \rightarrow w_i$ and dropping the error term to give the explicit formula for the 3-step method together with its estimated truncation error in the boxed equation. The formula employs the compact notation $f_i = f(t_i, w_i)$ and $w_i = w(t_i)$ for the functional evaluations at time $t_i = a + i \cdot h$; the truncation error term requires the third total time derivative $f^{(3)}(t, y)$ to be evaluated at a that point $(\xi_i, y(\xi_i))$ which maximizes the absolute error for a time ξ_i chosen in the interval $[t_i, t_{i+1}]$.

3.2.2 Implicit vs. Explicit Methods

Implicit *vs.* Explicit Methods

- **Example** $y' = f(t, y(t)) = -y + t + 1$; $y(0) = 1$; $t \in [0,1]$; Exact soln: $y = t + e^{-t}$
- **Set-up** $f_i = -w_i + t_i + 1$; $h = 0.1$; $t_i = 0.1i$

- **AB4 (Explicit)** $w_{i+1} = w_i + \dfrac{h}{24}[55 f_i - 59 f_{i-1} + 37 f_{i-2} - 9 f_{i-3}]$

$$w_{i+1} = \frac{1}{24}[18.5 w_i - 5.9 w_{i-1} - 3.7 w_{i-2} + .9 w_{i-3} + .24 i + 2.52]$$

- **AM3 (Implicit)** $w_{i+1} = w_i + \dfrac{h}{24}[9 f_{i+1} + 19 f_i - 5 f_{i-1} + f_{i-2}]$

w_{i+1} on both sides $\longrightarrow w_{i+1} = \dfrac{1}{24}[-.9 w_{i+1} + 22.1 w_i + .5 w_{i-1} - .1 w_{i-2} + .24 i + 2.52]$
- Solve for w_{i+1}

$$w_{i+1} = \frac{1}{24.9}[22.1 w_i + .5 w_{i-1} - .1 w_{i-2} + .24 i + 2.52]$$

- Implicit yields more accurate results, but must solve for w_{i+1}
- In general need Newton-Raphson for nonlinear eqn in w_{i+1}
- Instead: Use Explicit & Implicit together
  - AB4 Predicts $w^{(0)}_{i+1}$
  - Use as 1st guess for $w^{(0)}_{i+1}$ on RHS of AM3
  - AM3 Corrects solution yielding $w^{(1)}_{i+1}$

Here we numerically solve the IVP $\{y'=f(t,y(t))= -y+t+1 ; y(0)=1 ; t \in [0,1]\}$ with the exact solution $y=t +e^{-t}$ using both the explicit AB4 and the implicit AM3 methods. The set-up is the same for both methods, the slope function is evaluated as $f_i = f(t_i,w_i) = -w_i + t_i +1$ and for stepsize h=0.1 the indexed times are $t_i = 0 + i\,h = i\,h$; making these substitutions and collecting terms yields the results given on the slide for the two methods.

Note that for the implicit AM3 method the term w_{i+1} appears on both sides of the equation and in general this requires a Newton-Raphson iteration to find the step value w_{i+1}; however, because it appears linearly in the present case, we just collect the two terms on the LHS and divide by the coefficient as shown.

Although, not apparent from this example, the implicit method yields more accurate results, but of course requires several N-R iterations at each time step. The primary usage is to take the two together to form a Predictor-Corrector Method given on the next slide. The explicit AB4 predicts ahead to give a preliminary value for w_{i+1}, which serves as an initial "guess" on the RHS of the implicit AM3 equation; then direct calculation yields a corrected value for w_{i+1} without the need for N-R iterations.

Multi-Step Methods

3.2.3 Predictor-Corrector Methods – 4th Order

Predictor-Corrector Methods – 4th Order

1. Use RK4 to obtain "additional" start values w_1, w_2, w_3 from w_0

   $$w_1 = w_0 + \frac{1}{6}\left[k_1 + 2(k_2 + k_3) + k_4\right]$$

   $k_1 \equiv hf(t_0, w_0)$
   $k_2 \equiv hf(t_0 + \frac{h}{2}, w_0 + \frac{1}{2}k_1)$
   $k_3 \equiv hf(t_0 + \frac{h}{2}, w_0 + \frac{1}{2}k_2)$
   $k_4 \equiv hf(t_0 + h, w_0 + k_3)$

   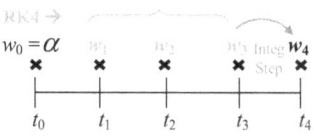

2. Predictor AB4 to find $w_4^{(0)}$

   $$w_4^{(0)} = w_3 + \frac{h}{24}\left[55f_3 - 59f_2 + 37f_1 - 9f_0\right]$$

   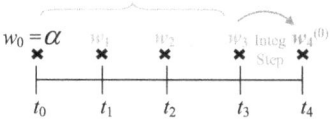

3. Corrector AM3 to find $w_4^{(1)}$

   $$w_4^{(1)} = w_3 + \frac{h}{24}\left[9f(t_4, w_4^{(0)}) + 19f_3 - 5f_2 + f_1\right]$$

   Would be implicit

   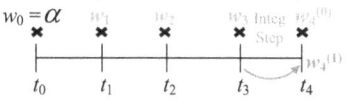

4. $w_4^{(1)}$ is the "final value" to be used in the next time step
   - Could continue iterations $w_4^{(2)}$, $w_4^{(3)}$ ...
   - In practice reduce stepsize instead
   - Method is quite stable (see stability discussion later)

The first three panels show the detailed implementation of the Predictor-Corrector Method which combines the explicit AB4 to predict a 1st guess for the implicit AM3 corrector.

Step #1 "starts" the AB4 method by producing 3 additional "initial" values using the "self-starting methods" RK4 formula to sequentially generate w_1, w_2, and w_3 as indicated by the values (green) under the brace in the figure of panel #1.

Step #2 applies the AB4 method by using the IVP initial condition $w_0 = \alpha$, together with the values w_1, w_2, and w_3 generated in step #1 as indicated by the values (green) under the brace in the figure of panel #2. to perform the forward integration step and yield $w_4^{(0)}$ with a superscripted "0" denoting that it is to be used as a start value for the implicit AM3 method in step #3 below.

Step #3 applies the AM3 method by using the three RK4-generated values $\{w_1, w_2, w_3\}$ together with the 0th iterate $w_4^{(0)}$ from the AB4 method as indicated by the values (green) under the brace in the figure of panel #3 and yields the 1st iterate $w_4^{(1)}$.

Step #3 can be repeated a number of times to yield the iterates $w_4^{(2)}$, $w_4^{(3)}$,... but in practice the iteration sequence converges slowly and in those instances in which the 1st iterate $w_4^{(1)}$ is not sufficient, the whole procedure needs to be restarted at step#1 with a smaller stepsize to obtain the desired accuracy. Stepsize control for the Predictor-Corrector method takes advantage of the different truncation error coefficients for these two 4th order methods and naturally allows estimation of the truncation error by setting $\Delta_{i+1} = |w_4^{(1)} - w_4^{(0)}|$ the difference between the implicit "AM3" and explicit "AB4" methods as detailed in the next slide. We shall also see that this method is quite stable to small perturbations.

3.2.4 Predictor-Corrector Error Control

Predictor-Corrector Error Control

- 4th Order Predictor AB4 $\quad y_{i+1} \cong w_{i+1}^{AB4} + \varepsilon_{i+1}^{AB4} \quad ; \quad \varepsilon_{i+1}^{AB4} = \frac{251}{720} y^{(5)}(\xi) h^5 \quad$ both order $n=4$

- 4th Order Corrector AM3 $\quad y_{i+1} \cong w_{i+1}^{AM3} + \varepsilon_{i+1}^{AM3} \quad ; \quad \varepsilon_{i+1}^{AM3} = -\frac{19}{720} y^{(5)}(\mu) h^5$

- Assume $y^{(5)}(\xi) \approx y^{(5)}(\mu)$

- Subtract to Estimate Trunc. Error for Corrector $\quad y^{(5)}(\mu) = \frac{8}{3h^5}\left[w_{i+1}^{AM3} - w_{i+1}^{AB4}\right] = \frac{8}{3h^5}\Delta_{i+1}$

- Calculate Stepsize Ratio using $\quad \tau[h] = \frac{\varepsilon_{i+1}^{AM3}}{h} = \left|-\frac{1}{h}\frac{19}{720}\left(\frac{8}{3h^5}\Delta_{i+1}\right)h^5\right| = \frac{19}{270}\frac{|\Delta_{i+1}|}{h}$

$$q = \left[\frac{\varepsilon_{TOL}}{\tau[h]}\right]^{1/n} = \left[\frac{\varepsilon_{TOL}}{\frac{19}{270}\frac{|\Delta_{i+1}|}{h}}\right]^{1/4} = \left[\frac{270 \cdot \varepsilon_{TOL} h}{19 \cdot |\Delta_{i+1}|}\right]^{1/4} \Longrightarrow \boxed{q = 1.63\left[\frac{\varepsilon_{TOL} h}{|\Delta_{i+1}|}\right]^{1/4}} \quad \text{"1.63" multiplier used instead of "1.94"}$$

- Changing Stepsize Expensive because need to re-compute start values 3 start values using RK4

  $q < 4 \quad$ Max Stepsize Ratio
  $\frac{\varepsilon_{TOL}}{10} < \frac{19}{270}\frac{|\Delta_{i+1}|}{h} < \varepsilon_{TOL} \quad$ No Stepsize change
  $h_{min} \leq h \leq h_{max} \quad$ min and max stepsize

Here we write down two expressions relating the exact i+1 step value y_{i+1} in terms of the two estimates $w^{AB4}_{i+1} = w^{(0)}_{i+1}$ and $w^{AM3}_{i+1} = w^{(1)}_{i+1}$ their associated errors. Since the LHS for both methods is the same, subtraction yields zero, and further assuming approximate equality of the two unknown 5th derivatives $y^{(5)}(\xi)$ and $y^{(5)}(\mu)$ we can obtain an estimate $y^{(5)}(\mu)$ as shown. Dividing the local truncation error by the stepsize "h" yields the cumulative truncation error $\tau[h] = (19/270) \Delta_{i+1}/h$ and finally using the expression for the stepsize ratio $q = (\varepsilon_{Tol}/\tau[h])^{1/n}$ for order n = 4 yields the results shown on the slide (except that the factor $(270/19)^{1/4} = 1.94$ is replaced by a more "conservative" value of 1.63 in the grey boxed result for q.

Also note that the second box gives some practical limits on the "range of stepsize ratios" by limiting the maximum ratio to be less than 4, and also by leaving the stepsize unchanged (q=1) whenever the cumulative error $\tau[h]$ is in the range $[0.1, 1]*\varepsilon_{Tol}$. Usually, there is also a minimum and maximum stepsize constraint as indicated in the second box.

These conditions are discovered by actual experience and are imposed in order to limit the number of "costly restarts" of the Predictor-Corrector Method since they involve the regeneration of new start values using the RK4 method for each change in stepsize. Usually stepsize control algorithms have rather complicated switching structures and other implementation issues, so on the next slide we discuss a few issues that may not be immediately apparent.

Multi-Step Methods

3.2.5 Example of Predictor-Corrector Method with Stepsize Control.

Example of Predictor-Corrector with Stepsize Control

IVP: $y' = y - t^2 + 1$; $t \in [0,2]$; $y(0) = 0.5$

$y_{exact} = (t+1)^2 - 0.5*\exp(t)$

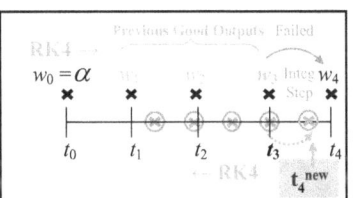

i	ti	wi	y_exact	h	τ[h]	|yi-wi|
0	0.0	0.5	0.5	0.25	1	0
1	0.1257017	0.7002318	0.7002323	0.125702	7.82E-05	5.20E-07
2	0.2514033	0.9230949	0.9230960	0.125702	7.82E-05	1.09E-06
3	0.3771050	1.1673877	1.1673894	0.125702	7.82E-05	1.72E-06
4	0.5028066	1.4317480	1.4317502	0.125702	7.82E-05	2.21E-06
5	0.6285083	1.7146306	1.7146334	0.125702	7.82E-05	2.80E-06
6	0.7542100	2.0142834	2.0142869	0.125702	7.82E-05	3.50E-06
7	0.8799116	2.3287200	2.3287244	0.125702	7.82E-05	4.35E-06
8	1.0056133	2.6556877	2.6556930	0.125702	7.82E-05	5.36E-06
9	1.1313149	2.9926319	2.9926385	0.125702	7.82E-05	6.56E-06
10	1.2570166	3.3366562	3.3366642	0.125702	7.82E-05	7.98E-06
11	1.3827183	3.6844761	3.6844857	0.125702	7.82E-05	9.67E-06
12	1.4857283	3.9697431	3.9697541	0.103010	1.11E-05	1.10E-05
13	1.5887383	4.2527705	4.2527830	0.103010	1.11E-05	1.26E-05
14	1.6917483	4.5310126	4.5310269	0.103010	1.11E-05	1.43E-05
15	1.7947583	4.8016476	4.8016639	0.103010	1.11E-05	1.63E-05
16	1.8977683	5.0615474	5.0615660	0.103010	1.11E-05	1.86E-05
17	2.0007783	5.3072447	5.3072658	0.103010	1.11E-05	2.11E-05

Variable Stepsize Control Parameters

$$h^{new} = q \cdot h^{old}$$

$$q = \left(\frac{\varepsilon_{TOL}}{2 \cdot \tau[h]} \right)^{1/4}$$

$$\tau[h] = \frac{19 \cdot |w_{i+1}^{pred} - w_{i+1}^{corr}|}{270 \cdot h}$$

$\varepsilon_{TOL} = 10^{-5}$

Stepsize Limits

$h_{min} = .01$
$h_{max} = .25$

Here is an output table for the same ODE as on Slides# 2-24, 2-28 but using the 4th order Adams Predictor-Corrector (AB4-AM3) method with stepsize control parameters $h_{max} = .25$, $h_{min} = .01$, and $\varepsilon_{TOL} = 10^{-5}$. The column layout is similar to previous tables with the iteration index i col#1, time t_i in col #2, numerical estimate w_i in col#3, exact solution y_{exact} in col#4, stepsize h in col #5, global truncation error $\tau[h]$ in col#6, and absolute error $|y_{exact}(t_i) - w_i|$ in col#7. The MatLab® script with inputs and outputs is given on Slides# 8-8, 8-9.

The quantity $\tau[h] = 19|\Delta_{i+1}|/(270h) = 19|w_{i+1}^{pred} - w_{i+1}^{corr}|/(270h)$ is an estimate of the global truncation error $\tau(h)$ obtained by combining the 4th order predictor and corrector solutions in the AB4-AM3 formulation (see Slide#3-7). It is used together with the specified error tolerance ε_{TOL} to compute the conservative estimate (extra factor of "2" in denominator) of the stepsize ratio $q = (\varepsilon_{TOL} / 2 \cdot \tau(h))^{1/4} = 1.63 (\varepsilon_{TOL} h / \Delta_{i+1})^{1/4}$. We see that the initial stepsize $h_0 = .25$ is immediately changed using $h_{new} = q*h_{old}$; values from the table yield $q = (\varepsilon_{TOL}/2\tau[h])^{1/4} = (10^{-5}/(2*7.822*10^{-5}))^{1/4} = .5028$, so $h_{new} = .5028*.25 = h_1$. Inspection of the table shows that the $h_1 = .1257$ maintains the error tolerance of 10^{-5} down the table until requires another stepsize reduction to $h_{12} = .10301$ at time t_{12}. This value is good for the remainder of the table to $t_{17} = 2.00077$; however, in order to maintain the error tolerance to times beyond those shown in the table (say to $t_i = 4$ and beyond), the minimum stepsize $h_{min} = .02$ may need to be lowered. Note also that since the table output values are not at uniform time intervals, the process of fitting an interpolation polynomial is more computationally intensive.

It is worthwhile studying the stepsize control switching structure in the code as some issues are not obvious at first blush. For example, a stepsize change requires a restart and instead of re-doing the *4 good output values* from the previous steps, *i.e.*, w_0, w_1, w_2, w_3, we keep those and do a **backward RK4** from the last good point w_3 (top right box on slide) to generate new start values ☻ that are then used for the next integration steps; these new points do not appear in the algorithm output and are discarded after they are no longer needed (*i.e.*, when a new stepsize is required). Note the new integration step to time t_4^{new} which corresponds to the adjusted step size.

3.3 Gragg's Extrapolation Technique (Romberg)

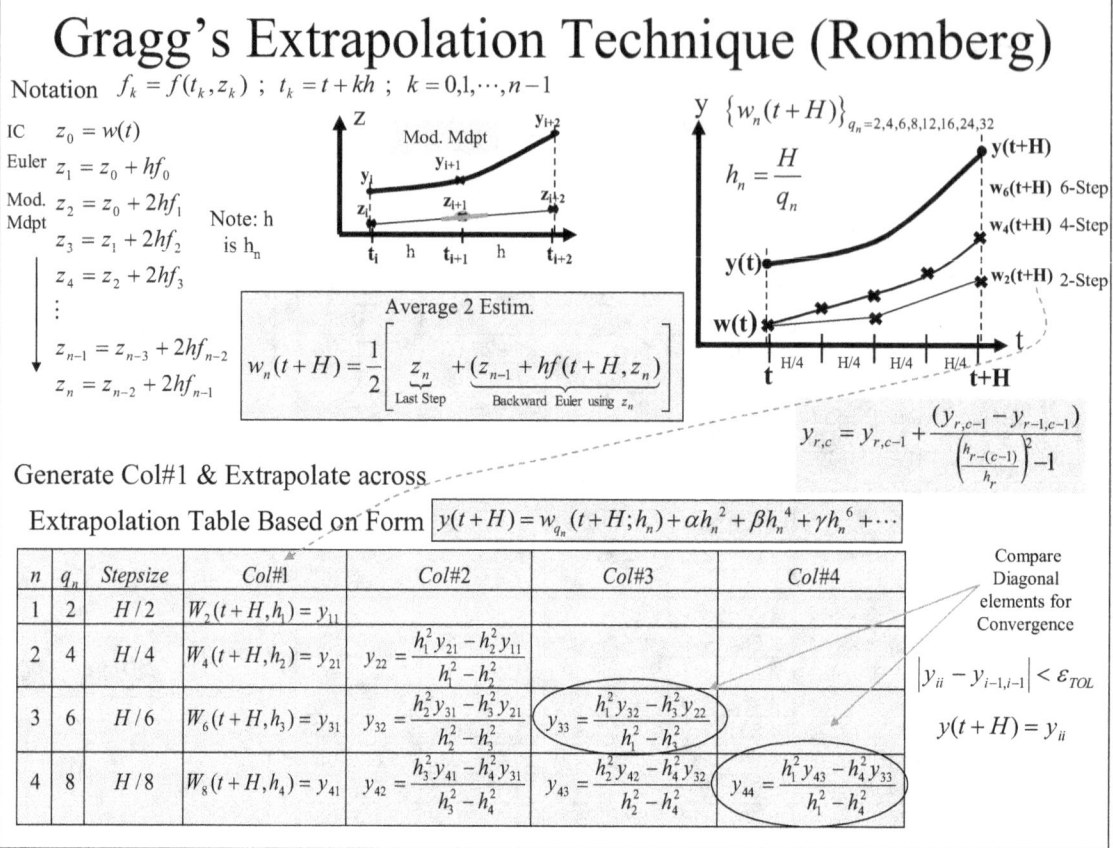

The solution to an IVP is usually desired at a fixed interval H and with a given accuracy ε_{Tol}. Recall the Romberg integration technique which combined composite trapezoidal panels with Richardson extrapolation to develop a "tableau" that achieved the desired accuracy. The known dependence of the composite trapezoidal algorithm on powers of "h" was key to this extrapolation technique. The Gragg Extrapolation Technique is an extension of these ideas to the solution of an IVP and once again the key is to develop a computational algorithm whose dependence upon powers of h is known, so that powers of h can be successively eliminated by extrapolation. The algorithm illustrated in the figure computes the output y(t+H) by first taking 2 steps of size H/2 to obtain an estimate $w_2(t+H)$ and then takes 4-steps of size H/4 to obtain a better estimate $w_4(t+H)$. The Gragg algorithm continues with better estimates by reducing stepsize (not always halving but following the sequence {H/2, H/4, H/6, H/8, H/12, H/16, H/24, H/32}) $h_n = H/q_n$ with ratio $q_n = \{2,4,6,8,12,16,24,32\}$ These values form col#1 of the table at the bottom of the slide and extrapolation to the right reduces truncation error by $O(h^2)$ in each successive column. *Gragg proved that the specific computational algorithm* described in the upper part of this slide yields estimates $w_n(t+H)$ that have the following error structure

$$y(t+H) = w_{qn}(t+H; h_n) + \alpha h_n^2 + \beta h_n^4 + \gamma h_n^6 + ...$$

where "h_n" in $w(t+H; h_n)$ emphasizes its dependence on stepsize, and the coefficients α, β, γ, are constants independent of "h". Note that the stepsizes are not halved down col#1 so the specific expressions for the extrapolation are spelled out in detail. Graggs algorithm for computing the "w"s uses a **2-step modified midpoint algorithm** which requires two start values: $\{z_0 = \alpha$ and $z_1 = \alpha_1\}$; thus we need an Euler "start-up" to obtain $z_1 = z_0 + hf(t_0, z_0)$ and then proceed with the

$$z_2 = z_0 + 2h\, f(t_1, z_1)\,;\ z_3 = z_1 + 2h\, f(t_2, z_2)\,;\ ...\,;\ z_n = z_{n-2} + 2h\, f(t_{n-1}, z_{n-1})$$

The estimates in col#1 of the table are given by averaging the final z_n and a backward Euler from the last point (t_n, z_n), viz., $w_{qn}(t+H; h_n) = \tfrac{1}{2}\,[z_n + \{z_{n-1} + h\, f(t_n, z_n)\}]$ [Note $t_n = t + H$]

Multi-Step Methods

3.3.1 Gragg's Extrapolation Example

Gragg's Extrapolation Example

- **Example** $y' = f(t, y(t)) = -y + t + 1$; $y(0) = 1$; $t \in [0,1]$; Exact soln: $y = t + e^{-t}$
- **Set-up** $f_i = -w_i + t_i + 1$; $z_0 = y(0) = 1$; $H = 0.25 = H_{max}$; $H_{min} = 0.02$

- **Sequence Stepsizes**

  $h_1 = H/2 = .1250$ 2–Steps → y_{11} $t_i = ih_1$
  $h_2 = H/4 = .0625$ 4–Steps → y_{21} $t_i = ih_2$
  $h_3 = H/6 = .04167$ 6–Steps → y_{31} $t_i = ih_3$
  $h_4 = H/8 = .03125$ 8–Steps → y_{41} $t_i = ih_4$

$w_2(t+H, h_1)$

- **2-Steps**

$h_1 = H/2 = .1250$

Step #1 (Euler) $z_1 = z_0 + h_1 f(t, z_0) = z_0 + h_1(-w_0 + t_0 + 1)$
$t_1 = \frac{h}{2} = .1250$ $z_1 = 1 + .1250(-w_0 + t_0 + 1) = 1 + .1250(-1 + 0 + 1) = 1$

Step #2 (Mod. Mdpt.) $z_2 = z_0 + 2h_1 f(t_1, z_1) = z_0 + 2h_1(-w_1 + t_1 + 1)$
$t_2 = 2\frac{h}{2} = .25$ $z_2 = 1 + 2(.1250)(-w_1 + t_1 + 1) = 1 + 2(.1250)(-1 + .1250 + 1) = 1.03125$

$$y_{11} = w_2(t+H, h_1) = \frac{1}{2}\left[z_2 + \{z_1 + h_1 f(t_2, z_2)\}\right] = \frac{1}{2}\left[z_2 + \{z_1 + h_1(-z_2 + t_2 + 1)\}\right]$$

$$= \frac{1}{2}[1.03125 + (1 + .1250(-1.03125 + 0 + .250 + 1))]$$

$\boxed{y_{11} = 1.0292969}$

- **Need to build extrapolation table**

$$y_{r,c} = y_{r,c-1} + \frac{(y_{r,c-1} - y_{r-1,c-1})}{\left(\frac{h_{r-(c-1)}}{h_r}\right)^2 - 1}$$

for $r, c = 2, 3, 4 \ldots$ & $r \geq c$

n	q_n	Stepsize	Col#1	Col#2	Col#3	Col#4
1	2	H/2	$W_2(t+H, h_1) = y_{11}$			
2	4	H/4	$W_4(t+H, h_2) = y_{21}$	$y_{22} = \frac{h_1^2 y_{21} - h_2^2 y_{11}}{h_1^2 - h_2^2}$		
3	6	H/6	$W_6(t+H, h_3) = y_{31}$	$y_{32} = \frac{h_2^2 y_{31} - h_3^2 y_{21}}{h_2^2 - h_3^2}$	$y_{33} = \frac{h_1^2 y_{32} - h_3^2 y_{22}}{h_1^2 - h_3^2}$	
4	8	H/8	$W_8(t+H, h_4) = y_{41}$	$y_{42} = \frac{h_3^2 y_{41} - h_4^2 y_{31}}{h_3^2 - h_4^2}$	$y_{43} = \frac{h_2^2 y_{42} - h_4^2 y_{32}}{h_2^2 - h_4^2}$	$y_{44} = \frac{h_1^2 y_{43} - h_4^2 y_{33}}{h_1^2 - h_4^2}$

In setting up a Gragg Extrapolation, the sequence of stepsize ratios only runs through the set of eight values $q_n = \{2,4,6,8,12,16,24,32\}$ generating the eight elements in col#1 $w_{qn}(t+H; h_n)$ for $n=1,2,\ldots 8$; if the resulting extrapolation table for the current output H-step fails to achieve the desired accuracy ε_{Tol}, then the output H-value is "halved" and the whole procedure is repeated; halving of H continues until it becomes less than $H_{min} = .02$ at which point the algorithm is stopped. The best way to understand the details is to program it in a language of your choice (not a spreadsheet!) and see what numerical recipe you come up with.

We apply Gragg's Extrapolation to the IVP $\{y'=f(t,y(t))= -y+t+1$; $y(0)=1$; $t\in[0,1]$ with $H_{max}=.25$ and $H_{min}=.02\}$ with exact soln $y=t+e^{-t}$. The slope function is evaluated as $f_i = f(t_i,w_i) = -w_i + t_i + 1$ and the output interval is set initially to be a maximum value $H_{max}=.25$ which is maintained so long as the extrapolated output is within the desired tolerance ε_{Tol}. Here we illustrate the set up and computational flow for only the first member in col#1 "$w_{qn}(t+H; h_n)$" => $w_2(1+.25; .25/2)$ which is trivial enough to fit on the slide. We need to generate the remaining members $w_4(1+.25; .25/4)$, $w_6(1+.25; .25/6)$, etc.; this sequence can stop when the difference between successive diagonal values is less than the desired accuracy, ε_{Tol} at which point we output the one H-step value the last diagonal element of the matrix, say, $w[1.25] = y_{44}[1.25]$. The remainder of this calculation is found in the MatLab® script with inputs and outputs on Slides# 8-10, 8-11. Our standard ODE example used for method comparisons (on Slides# 2-24, 2-28, 3-8) is given for the Gragg Extrapolation method on the next slide.

3.3.2 Gragg Extrapolation Example Detailed Output Tables

Gragg Extrapolation Example Detailed Output Tables

IVP: $y' = y - t^2 + 1$; $t \in [0, 2]$; $y(0) = 0.5$

$y_{exact} = (t+1)^2 - 0.5 * \exp(t)$

Extrapolation Stepsize Sequence
$q_i = [2\ 4\ 6\ 8\ 12\ 16\ 24\ 32]$
$h_i = H/q_i$;
$H = 0.25$; $H_{min} = .02$; $e_{TOL} = 10^{-5}$

i	t_i	w_i	y_exact	H	$|y_i - w_i|$	k_extrap
0	0	0.5	0.5	0.25	0	0
1	0.25	0.920487291656	0.920487291656	0.25	1.55E-15	5
2	0.5	1.425639364650	1.425639364650	0.25	4.00E-15	5
3	0.75	2.003999991694	2.003999991694	0.25	5.33E-15	5
4	1	2.640859085770	2.640859085770	0.25	7.55E-15	5
5	1.25	3.317328521270	3.317328521269	0.25	5.08E-13	4
6	1.5	4.009155464832	4.009155464831	0.25	6.54E-13	5
7	1.75	4.685198661998	4.685198661997	0.25	8.47E-13	5
8	2	5.305471950536	5.305471950535	0.25	1.10E-12	5
9	2.25	5.818632081822	5.818632081821	0.25	1.41E-12	5

i=1
0.918701	0	0	0	0
0.920038	0.920484	0	0	0
0.920287	0.920487	0.920487	0	0
0.920375	0.920487	0.920487	0.920487	0
0.920437	0.920487	0.920487	0.920487	0.920487

i=2
1.42397	0	0	0	0
1.42522	1.425636	0	0	0
1.425453	1.425639	1.425639	0	0
1.425534	1.425639	1.425639	1.425639	0
1.425593	1.425639	1.425639	1.425639	1.425639

i=3
2.002481	0	0	0	0
2.003618	2.003997	0	0	0
2.00383	2.004	2.004	0	0
2.003904	2.004	2.004	2.004	0
2.003958	2.004	2.004	2.004	2.004

i=4
2.639532	0	0	0	0
2.640526	2.640857	0	0	0
2.640711	2.640859	2.640859	0	0
2.640776	2.640859	2.640859	2.640859	0
2.640822	2.640859	2.640859	2.640859	2.640859

i=5
3.316249	0	0	0
3.317058	3.317328	0	0
3.317208	3.317328	3.317329	0
3.317261	3.317328	3.317329	3.317329

i=6
4.008393	0	0	0	0
4.008965	4.009156	0	0	0
4.009071	4.009156	4.009155	0	0
4.009108	4.009155	4.009155	4.009155	0
4.009134	4.009155	4.009155	4.009155	4.009155

i=7
4.684844	0	0	0	0
4.685112	4.685201	0	0	0
4.68516	4.685199	4.685199	0	0
4.685177	4.685199	4.685199	4.685199	0
4.685189	4.685199	4.685199	4.685199	4.685199

i=8
5.305641	0	0	0	0
5.305517	5.305476	0	0	0
5.305492	5.305472	5.305472	0	0
5.305484	5.305472	5.305472	5.305472	0
5.305477	5.305472	5.305472	5.305472	5.305472

i=9
5.819473	0	0	0	0
5.818848	5.818639	0	0	0
5.818728	5.818633	5.818632	0	0
5.818686	5.818632	5.818632	5.818632	0
5.818656	5.818632	5.818632	5.818632	5.818632

Here is a complete set of detailed output tables for the Gragg extrapolation method with error control and a uniform output stepsize H (=.25). Note that error control here is done *via* an extrapolation table which chooses a particular sequence of intermediate stepsizes; the output stepsize however is fixed at H. The main output table should be compared with the results for the same ODE previously implemented for other methods on Slides# 2-24 (Euler, RK4), 2-28 (RKF45), and 3-8 (Pred-Corr). The MatLab® script with inputs and outputs is given on Slides# 8-8, 8-9.

The parameters H = .25, H_{min}=.02, and ε_{TOL} =10^{-5} are similar to those used in the other methods, except here the desired stepsize for the output table is fixed at H=.25 and the H_{min} parameter allows the algorithm to *halve the fixed stepsize* to H/2 for cases in which the *extrapolation stepsize sequence* ends without meeting the tolerance stopping criterion ε_{TOL} =10^{-5} . This of course requires two H/2 steps to generate the new output value; the H_{min} parameter limits the number of times H may be halved before the algorithm quits with an exception error. The column layout is similar to previous tables with the iteration index *i* in col#1, time t_i in col #2, numerical estimate w_i in col#3, exact solution y_{exact} in col#4, fixed stepsize H in col #5, absolute error $|y_{exact}(t_i) - w_i|$ in col#6, and the #rows k_extrap in each extrapolation table. The complete set of extrapolation tables for each row of the main output table i=1,2,…,9 is given with the final extrapolation value satisfying the stopping criterion comparing diagonal elements $|w_{i+1} - w_i| < \varepsilon_{TOL}$ =10^{-5} emphasized in grey. We note that for this particular ODE all of these extrapolation tables have k_extrap = 5, except for i=5 which satisfies the stopping criterion with k_extrap = 4; also note that although ε_{TOL} is set at 10^{-5} the actual error $|y_{exact}(t_i) - w_i|$ in col#6 is significantly smaller at 10^{-12} to 10^{-15}. Finally, we note that the Gragg extrapolation output table with its *uniform output stepsize* H is ideally suited to polynomial interpolation applications.

4 Systems and Higher Order IVPs

Systems and Higher Order IVPs

4.1 Coupled Systems of 1st Order Differential Equations

Coupled Systems of 1st Order Differential Equations

System of Two 1st Order Eqns:
$$\frac{du_1}{dt} = f_1(t, u_1, u_2) \; ; \; u_1(a) = \alpha_1$$

Coupled

$$\frac{du_2}{dt} = f_2(t, u_1, u_2) \; ; \; u_2(a) = \alpha_2$$

2^d Lifschitz Condition on Domain D:
$$|f_k(t, u_1, u_2) - f_k(t, v_1, v_2)| \leq L\{|u_1 - v_1| + |u_2 - v_2|\} \quad k = 1, 2$$

$D: t \in [a, b] \; ; \; u_1, u_2 \in (-\infty, +\infty)$

Equivalent Derivative Conditions:
$$\left|\frac{\partial f_k(t, u_1, u_2)}{\partial u_1}\right| \leq L \qquad \left|\frac{\partial f_k(t, u_1, u_2)}{\partial u_2}\right| \leq L$$

Theorem: Unique Solution if f_1, f_2 Continuous & Satisfy Lifschitz Conditions.

Vectorize RK4 Eqns:
$$\vec{w} = \begin{bmatrix} w_1 \\ w_2 \end{bmatrix} \; ; \; \vec{f} = \begin{bmatrix} f_1 \\ f_2 \end{bmatrix} \; ; \; \vec{\alpha} = \begin{bmatrix} \alpha_1 \\ \alpha_2 \end{bmatrix}$$

$$\vec{k}_1 = h\vec{f}(t_0, \vec{w}_0)$$

$$\vec{k}_2 = h\vec{f}(t_0 + \frac{h}{2}, \vec{w}_0 + \frac{\vec{k}_1}{2})$$

$$\vec{w}_{i+1} = \vec{w}_i + \frac{1}{6}[\vec{k}_1 + 2(\vec{k}_2 + \vec{k}_3) + \vec{k}_4]$$

$$\vec{k}_3 = h\vec{f}(t_0 + \frac{h}{2}, \vec{w}_0 + \frac{\vec{k}_2}{2})$$

$$\vec{k}_4 = h\vec{f}(t_0 + h, \vec{w}_0 + \vec{k}_3)$$

The Fundamental IVP is now extended to a coupled system of 1st order DEs. Specifically, for n=2 we have written down two 1st order DEs with two initial conditions and the coupling between them is evident from the form of the two slope functions $f_1(t, u_1, u_2)$ and $f_2(t, u_1, u_2)$. In a fashion analogous to the original 1^d Lipschitz conditions on f(t,y) in the domain D$\{t \in [a,b], y \in (-\infty, +\infty)\}$ we now have a 2^d Lipschitz conditions in the new Domain D$\{t \in [a,b], u, v \in (-\infty, +\infty)\}$ written as shown:
$|f_1(t, u_1, u_2) - f_1(t, v_1, v_2)| < L \{|u_1 - v_1| + |u_2 - v_2|\}$ and likewise for the second function $f_2(t, u_1, u_2)$. There is also a theorem relating the above Lipschitz conditions to the partials with respect to u_1 and u_2 $|\partial/\partial u_1 f_1(t, u_1, u_2)| < L$ and $|\partial/\partial u_2 f_1(t, u_1, u_2)| < L$ and likewise for the second function $f_2(t, u_1, u_2)$. Finally, there exists a unique solution to the coupled system if f_1 and f_2 are continuous and satisfy the Liftshitz conditions over the restricted domain D specified by the IVP.

The RK4 method for a pair of coupled DEs is displayed in vector form, in terms of the 2-vectors **w**, **f**, { **k$_1$, k$_2$, k$_3$, k$_4$**}, and **α** (bolded) and they clearly look like the original RK4 equations except for the vector character of all the variables. Although they are identical in structure, there is a layer of complexity that arises when the equations explicitly written out in component form and indexed for a computer algorithm. In this respect, note that when a vector occurs inside a function this means that the function has all vector components as variables, e.g., $f_1(t, \mathbf{u}) = f_1(t, u_1, u_2)$; this is especially important to recognize when extended to a system of n 1st order DEs where now $f_1(t, \mathbf{u}) = f_1(t, u_1, u_2, u_3, ..., u_n)$. Also note that the independent variable "t" is not a vector and always has one component. An explicit example is given in the next slide.

Systems and Higher Order IVPs

4.1.1 Kirchoff System for Currents - Example

Kirchoff System for Currents - Example

System of Eqns:

$$\frac{dI_1}{dt} = f_1(t, I_1, I_2) = -4I_1 + 3I_2 + 6 \quad ; \quad I_1(0) = 0$$

$$\frac{dI_2}{dt} = f_2(t, I_1, I_2) = -2.4I_1 + 1.6I_2 + 3.6 \quad ; \quad I_2(0) = 0$$

$$\vec{w} = \begin{bmatrix} I_1 \\ I_2 \end{bmatrix} = \begin{bmatrix} w_1 \\ w_2 \end{bmatrix} \qquad \vec{f} = \begin{bmatrix} -4w_1 + 3w_2 + 6 \\ -2.w_1 + 1.6w_2 + 3.6 \end{bmatrix}$$

$$\vec{w}_0 = \begin{bmatrix} (w_0)_1 \\ (w_0)_2 \end{bmatrix} = \vec{\alpha} = \begin{bmatrix} 0 \\ 0 \end{bmatrix} \qquad h = 0.1$$

$$t_i = ih = 0.1 \cdot i \quad for \quad i = 0, 1, 2, \cdots$$

RK4 Step: $t_0 \to t_1$

$$\vec{k}_1 = h\vec{f}(t_0, \vec{w}_0) = 0.1 \begin{bmatrix} -4(w_0)_1 + 3(w_0)_2 + 6 \\ -2.4(w_0)_1 + 1.6(w_0)_2 + 3.6 \end{bmatrix} = \begin{bmatrix} 0.6 \\ 0.36 \end{bmatrix}$$

Exact Soln:

$$I_1(t) = -3.375e^{-2t} + 1.875e^{-0.4t} + 1.5$$

$$I_2(t) = -2.25e^{-2t} + 2.25e^{-0.4t}$$

$$\vec{k}_2 = h\vec{f}(t_0 + \frac{h}{2}, \vec{w}_0 + \frac{\vec{k}_1}{2}) = 0.1 \begin{bmatrix} -4\{(w_0)_1 + \frac{.6}{2}\} + 3\{(w_0)_2 + \frac{.36}{2}\} + 6 \\ -2.4\{(w_0)_1 + \frac{.6}{2}\} + 1.6\{(w_0)_2 + \frac{.36}{2}\} + 3.6 \end{bmatrix} = \begin{bmatrix} .534 \\ .3168 \end{bmatrix}$$

$$\vec{k}_3 = h\vec{f}(t_0 + \frac{h}{2}, \vec{w}_0 + \frac{\vec{k}_2}{2}) = 0.1 \begin{bmatrix} -4\{(w_0)_1 + \frac{.534}{2}\} + 3\{(w_0)_2 + \frac{.3168}{2}\} + 6 \\ -2.4\{(w_0)_1 + \frac{.534}{2}\} + 1.6\{(w_0)_2 + \frac{.3168}{2}\} + 3.6 \end{bmatrix} = \begin{bmatrix} .54072 \\ .321264 \end{bmatrix}$$

$$\vec{k}_4 = h\vec{f}(t_0 + h, \vec{w}_0 + \vec{k}_3) = 0.1 \begin{bmatrix} -4\{(w_0)_1 + .54072\} + 3\{(w_0)_2 + .321264\} + 6 \\ -2.4\{(w_0)_1 + .54072\} + 1.6\{(w_0)_2 + .321264\} + 3.6 \end{bmatrix} = \begin{bmatrix} .4800912 \\ .28162944 \end{bmatrix}$$

$$\vec{w}_1 = \vec{w}_0 + \frac{1}{6}[\vec{k}_1 + 2(\vec{k}_2 + \vec{k}_3) + \vec{k}_4] = \begin{bmatrix} 0 \\ 0 \end{bmatrix} + \frac{1}{6}\left\{\begin{bmatrix} 0.6 \\ 0.36 \end{bmatrix} + 2\left(\begin{bmatrix} .534 \\ .3168 \end{bmatrix} + \begin{bmatrix} .54072 \\ .321264 \end{bmatrix}\right) + \begin{bmatrix} .4800912 \\ .28162944 \end{bmatrix}\right\}$$

5 Digit Accuracy:

$$\vec{w}_1 = \begin{bmatrix} .5382552 \\ .31962624 \end{bmatrix} \qquad \begin{array}{l} I_1(0.1) = .53826390677 \\ I_2(0.1) = .31963204366 \end{array}$$

This Kirchoff equations are a pair of coupled DEs which determine the currents I_1, I_2 in a "two loop" electrical circuit are re-written as vectorized difference equations for the 2-vector $\mathbf{w} = [w_1, w_2]^T$ on the top right of the slide. The RK4 technique is applied in "2-vector form" computing $\mathbf{k}_1 = [k_{1,1}, k_{1,2}]^T$ as a column vector first using the column vector function $\mathbf{f} = [f_1, f_2]^T$, which has two different functions in its components and continuing in this fashion to find $\mathbf{k}_2$, $\mathbf{k}_3$, $\mathbf{k}_4$ and then using the RK4 update equation to find the two current components at time t_1, *i.e.*, $\mathbf{w}_1 = [.53826, .31963]^T$ which agrees with the exact solution to 5 significant digits. This process is continued until the full range of desired outputs is computed; clearly the extension to complicated multi-loop circuits is easily written down and computed in exactly the same manner; the only caution it to program with special attention given to the indexing scheme.

Systems and Higher Order IVPs

4.2 Higher Order Differential Equations

Higher Order Differential Equations

3rd Order ODE:
$$y''' = f(t, y, y', y'') \quad ; \quad y(0) = \alpha_1 \ ; \ y'(0) = \alpha_2 \ ; \ y''(0) = \alpha_3$$

Vectorize:
$$\vec{y} \equiv \begin{bmatrix} y_1 \\ y_2 \\ y_3 \end{bmatrix} = \begin{bmatrix} y(t) \\ y'(t) \\ y''(t) \end{bmatrix} \qquad \begin{array}{l} y_1' = y_2 \ ; \ y_2' = y_3 \ ; \\ \\ y_3' = f(t, y, y', y'') = f(t, y_1, y_2, y_3) = f(t, \vec{y}) \end{array}$$

Differentiate:
$$\vec{y}' = \begin{bmatrix} y_1' \\ y_2' \\ y_3' \end{bmatrix} = \begin{bmatrix} y_2 \\ y_3 \\ f(t, \vec{y}) \end{bmatrix} \equiv \begin{bmatrix} f_1(t, \vec{y}) \\ f_2(t, \vec{y}) \\ f_3(t, \vec{y}) \end{bmatrix}$$

Equivalent to System of Three 1st Order ODEs:
$$\vec{y}' = \vec{f}(t, \vec{y}) \ ; \ \vec{y}(0) = \vec{\alpha} \qquad \vec{f}(t, \vec{y}) \equiv \begin{bmatrix} f_1(t, \vec{y}) \\ f_2(t, \vec{y}) \\ f_3(t, \vec{y}) \end{bmatrix} \qquad \vec{\alpha} \equiv \begin{bmatrix} \alpha_1 \\ \alpha_2 \\ \alpha_3 \end{bmatrix}$$

Higher order differential equations such as the 3rd order one shown on this slide are vectorized by defining the 3-vector components as the original variable $y_1 = y$ and its two derivatives $y_2 = y'$ and $y_3 = y''$. Differentiating the 3-vector $\mathbf{y'} = [y_1', y_2', y_3']^T = [y', y'', y''']^T$ and using the differential equation and definitions of components, we find $\mathbf{y'} = [y_2, y_3, f(t, y_1, y_2, y_3)]^T$; finally identifying components of the vector $\mathbf{y'}$ with the components of the vector function $\mathbf{f}$ yields
$$\mathbf{f} = [f_1(t, y_1, y_2, y_3)], f_2(t, y_1, y_2, y_3)], f_3(t, y_1, y_2, y_3)]^T$$
Thus we are led to a vector formulation for the 3rd order DE which is equivalent to a system of three coupled 1st order DEs using the vector quantities $\mathbf{y}$, $\mathbf{f}$, and $\boldsymbol{\alpha}$ defined in the boxed set of equations. Thus in general nth order DEs reduce to solutions of n coupled 1st order DEs and are solved in the same manner as the Kirckoff example of the last slide.

Systems and Higher Order IVPs

4.2.1 Setup for a 2nd Order ODE

Setup for a 2nd Order ODE

2nd Order IVP: $y'' - 2y' + 2y = e^{2t} \sin t$; $t \in [0,1]$; $y(0) = -0.4$; $y'(0) = -0.6$

$$y'' = -2y + 2y' + e^t \sin t = f(t, y, y')$$

Define y_1, y_2
$y_1 = y$
$y_2 = y_1'$

→

Vectorize
$$\vec{y} = \begin{bmatrix} y_1 \\ y_2 \end{bmatrix} = \begin{bmatrix} y \\ y' \end{bmatrix}$$

→

Differentiate
$$\vec{y}' = \begin{bmatrix} y_1' \\ y_2' \end{bmatrix} = \begin{bmatrix} y' \\ y'' \end{bmatrix} = \begin{bmatrix} y_2 \\ f(t, \vec{y}) \end{bmatrix}$$

Rewrite DE in terms of y_1, y_2

$$y_2 = y_1' = +0 \cdot y_1 + 1 \cdot y_2 + 0$$
$$y_2' = -2 \cdot y_1 + 2 \cdot y_2 + e^t \cdot \sin t$$

Form Vector-Matrix Eqns for Equivalent Coupled System.

$$\begin{bmatrix} y_1' \\ y_2' \end{bmatrix} = \begin{bmatrix} 0 & 1 \\ -2 & 2 \end{bmatrix} \begin{bmatrix} y_1 \\ y_2 \end{bmatrix} + \begin{bmatrix} 0 \\ e^t \sin t \end{bmatrix} \quad ; \quad \vec{y}(o) = \begin{bmatrix} y_1(0) \\ y_2(0) \end{bmatrix} = \begin{bmatrix} -0.4 \\ -0.6 \end{bmatrix}$$

For a typical 2nd order DE with two initial conditions we have the IVP
$$y'' - 2y' + 2y = e^t \sin(t) \; ; \; t \in [0,1] \; ; \; y(0) = -0.4; \; y'(0) = -0.6$$
The DE is put into the standard form $y'' = f(t, y, y') = +2y' - 2y + e^t \sin(t)$, thus defining the function $f(t, y, y')$. Further defining the 2-vector components as $y_1 = y$ and $y_2 = y'$, we generate the equivalent system of coupled DEs in a form that is ready for vector computations. Note that we have introduced a 2x2 matrix which multiplies the 2-vector $\mathbf{y} = [y_1, y_2]^T$ which may prove convenient in formulating the computations as a coupled system of linear differential equations $\mathbf{y}' = A\mathbf{y} + \mathbf{b}$; this form is equivalent to writing down the two equations individually as in the Kirchoff example. Application of RK4 algorithm to the difference equation for this problem is the same as for the Kirchoff example where we wrote down the vectorized equations in terms of the 2-vectors $\mathbf{w}$, $\mathbf{f}$, { k_1, k_2, k_3, k_4 }, and α

4.2.2 2nd Order ODE Example Using RK4

2nd Order ODE Example Using RK4

2nd Order IVP: $y''-3y'+2y=t$; $t \in [0,3]$; $y(0)=0$; $y'(0)=0.5$

i	ti	w1i	w2i	y	yp	|yi-w1i|	|ypi-w2i|
0	0	0	0.5	0	0.5	0	0
1	0.1	0.058294	0.67434	0.058296	0.67435	1.94E-06	4.01E-06
2	0.2	0.13676	0.90562	0.13676	0.90563	4.77E-06	9.83E-06
3	0.3	0.24179	1.2084	0.2418	1.2084	8.79E-06	1.81E-05
4	0.4	0.3814	1.6005	0.38142	1.6006	1.44E-05	2.95E-05
5	0.5	0.56561	2.1043	0.56563	2.1043	2.21E-05	4.51E-05
6	0.6	0.80688	2.7469	0.80691	2.747	3.25E-05	6.62E-05
7	0.7	1.1207	3.5621	1.1208	3.5622	4.65E-05	9.45E-05
8	0.8	1.5264	4.5911	1.5265	4.5912	6.51E-05	0.000132
9	0.9	2.0477	5.8849	2.0478	5.8851	8.97E-05	0.000182
10	1	2.7142	7.5059	2.7144	7.5062	0.000122	0.000247
11	1.1	3.5623	9.5309	3.5625	9.5313	0.000164	0.000332
12	1.2	4.637	12.054	4.6372	12.055	0.000219	0.000443
13	1.3	5.9936	15.191	5.9939	15.192	0.000291	0.000587
14	1.4	7.7003	19.083	7.7007	19.084	0.000383	0.000773
15	1.5	9.8411	23.905	9.8416	23.906	0.000503	0.001013
16	1.6	12.519	29.868	12.52	29.869	0.000656	0.00132
17	1.7	15.861	37.234	15.862	37.235	0.000852	0.001715
18	1.8	20.023	46.321	20.024	46.323	0.001103	0.002219
19	1.9	25.196	57.52	25.197	57.523	0.001424	0.002862
20	2	31.613	71.31	31.615	71.314	0.001832	0.003682
21	2.1	39.563	88.276	39.565	88.28	0.002352	0.004724
22	2.2	49.398	109.13	49.401	109.14	0.003012	0.006047
23	2.3	61.548	134.76	61.552	134.77	0.003849	0.007724
24	2.4	76.543	166.22	76.548	166.23	0.004909	0.009848
25	2.5	95.03	204.83	95.036	204.85	0.006249	0.012533
26	2.6	117.8	252.2	117.81	252.21	0.007942	0.015924
27	2.7	145.83	310.27	145.84	310.29	0.010078	0.020202
28	2.8	180.29	381.45	180.3	381.47	0.012771	0.025595
29	2.9	222.65	468.66	222.66	468.69	0.016162	0.032384
30	3	274.67	575.47	274.69	575.51	0.020428	0.040925

$$y_{exact} = (t+1)^2 - .5e^t$$

Equivalent System of Equations

Set: $y_1 = y$; $y_2 = y'$

$y_1' = f_1(t, y_1, y_2) = y_2$

$y_2' = f_2(t, y_1, y_2) = t - 2 \cdot y_1 + 3 \cdot y_2$

$y_1(0) = 0$; $y_2(0) = 0.5$

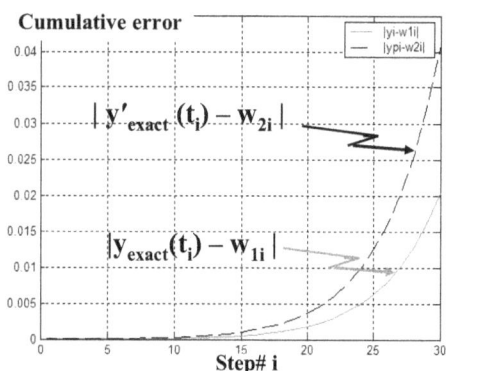

Cumulative error

Here is a detailed output table for a 2nd order ODE using the *vectorized* RK4 method for systems of equations *without error control*. Note that error control can easily be invoked here by either of the two methods discussed on Slide#2-25 resulting respectively in RK4 with halving or RKF45 using the 4th and 5th order solutions. The MatLab® script with inputs and outputs is given on Slides# 8-12, 8-13. The column layout is similar to previous tables except the coupled system solution yields pairs of estimates for the function and its derivative; in this case we have the iteration index *i* in col#1, time t_i in col #2, numerical estimates w_{1i} and w_{2i} in cols#3,4, exact solution y_{exact} and its derivative y'_{exact} in cols#5,6, and finally the absolute errors $|y_{exact}(t_i) - w_{1i}|$ and $|y'_{exact}(t_i) - w_{2i}|$ in cols#7,8. A typical reduction of a higher order IVP to that for system of coupled equations is given on the previous Slide#4-5. The specific transformation of variables leading to the equivalent system of two equations for the variables $y_1 = y$ and $y_2 = y'$ of this example is given in the grey boxed equations on the slide.

As we have seen before (Slide#2-28), without error control, the error in both the solution and its derivative degrade as the cumulative truncation error builds with each step; thus in the last two columns the absolute errors $|y_{exact}(t_i) - w_{1i}|$ and $|y'_{exact}(t_i) - w_{2i}|$ start at 10^{-6} and after 30 steps of h=0.1 the errors are quite large of order 10^{-2}. This error growth as a function of step index i is illustrated in the plot at the bottom right, where the solid curve (red) represents the solution to the IVP w_{1i} and the dashed curve (black) is for its derivative w_{2i}. Clearly, for sustained accuracy over a long run of values, we need to employ some form of error control.

5 Special Numerical Considerations for IVP Techniques

Special Numerical Considerations for IVP Techniques

5.1 Consistency, Convergence, Stability

	Consistency (Method)	**Convergence** (Soln)	**Stability** (Computation)
• IVP	$y' \equiv \dfrac{dy}{dt} = f(t, y(t))\quad t \in [a,b]\quad IC: y(t=a) = \alpha$		
• Sampled Slope Format	$y_{i+1} = y_i + h\underbrace{\phi^{(n)}(t_i, y_i; h)}_{\text{method of order "n"}} + \underbrace{\varepsilon_{i+1}^{trunc}[h]}_{\text{Method Trunc. Error}}$		$\phi^{(n)}(t_i, y_i; h) = \sum_i a_i m_i *$ Quadrature: weighted sum of slopes
• Transition to Difference Eqn (Replace $y_i \to w_i$ & Drop $\varepsilon_{i+1}^{trunc}[h]$)	$w_{i+1} = w_i + \underbrace{h\phi(t_i, w_i; h)}_{\text{method}}$		$\phi^{(n)}(t_i, w_i; h) = \sum_i a_i m_i$

1. Consistency of "Method" *Difference* Eqn → *Differential* Eqn
(Bkwards: Method → DE)

1) Arb. slope method may not be valid
2) Taylor Derived always consistent

$w_i \to y_i$

$y_{i+1} = y_i + h\underbrace{\phi^{(n)}(t_i, y_i; h)}_{\text{method with } y_i}$

Local Trunc $\varepsilon^{trunc} = h*\tau[h]$

$\varepsilon^{trunc} = h*\tau[h]$: (exact soln to DE y_{i+1}) − (DfE "method" value $y_i + h\phi$)

$\lim_{h \to 0} \tau_{i+1}^{trunc}[h] = \lim_{h \to 0} \dfrac{y_{i+1} - [y_i + h\phi(t_i, y_i)]}{h} = \lim_{h \to 0} \left| \dfrac{y_{i+1} - y_i}{h} - \phi^{(n)}(t_i, y_i; h) \right| = 0$

2. Convergence of "Solution" Soln to *Difference* Eqn → Soln to *Differential* Eqn

$\lim_{h \to 0} \phi(t_i, y_i; h) = f(t_i, y_i)$ $\qquad \lim_{h \to 0} |y_{i+1} - w_{i+1}| = 0$

3. Stability of Solution to R.O. Error: Perturb IC $\quad \alpha \to \alpha + \varepsilon \quad$ *Still Converges*

The integration over a single step h of a differential equation (DE) is formally written down in the "sampled slope format" as a relation between y_{i+1} and y_i; this relationship is exact because of the addition of the truncation error ε_{i+1} associated with the "method" at the time t_{i+1} on the RHS. The "n^{th} order slope method" is defined by the function $\phi^{(n)}(t_i, y_i; h)$ which is a linear combination of a set of **true slopes** m_i*. The transition to a difference equation (DfE) involves two approximations, (i) dropping the truncation error and (ii) replacing y_i by w_i; it yields the difference equation

$$w_{i+1} = w_i + h\phi^{(n)}(t_i, w_i; h)$$

where the function $\phi^{(n)}(t_i, w_i; h)$ is now a linear combination of the **approximate slopes** m_i.
If we are presented with a method $\phi^{(n)}(t_i, w_i; h)$ of unknown origin, we would like to be certain of three things, as follows:
1) Consistency of the difference equation (DfE) with the original **differential equation (DE)**: are we solving the right problem? For the DfE to be equivalent to the DE we must show that the quantity $\tau[h]$ defined as the difference between $(y_{i+1} - y_i)/h$ and the solution method slope function $\phi^{(n)}(t_i, y_i; h)$ approaches zero as as $h \to 0$ viz.,

$$\lim_{h \to 0} \tau[h] = | (y_{i+1} - y_i)/h - \phi^{(n)}(t_i, y_i; h) | \to 0 .$$

(This is accomplished by making n^{th} order Taylor expansions of y_{i+1} and terms in the expression $\phi^{(n)}(t_i, y_i; h)$ about t_i.)
2) Convergence of the solution: does the **solution of the difference equation converge** to that of the original DE:

$$| y_{i+1} - w_{i+1} | \to 0 \text{ as } h \to 0.$$

3) Stability of the solution to small perturbations; that is, do we still have convergence when we perturb the initial conditions $\alpha \to \alpha + \varepsilon$.

Note that the quantity $\tau[h]$ in (1) is also called the "local truncation error" and is equal to h times what we have previously called the local truncation error (see figure). The quantity $\tau[h] = \varepsilon_{i+1}/h$ is used to test how well the solution to the difference equation (DfE) tracks the exact solution y_{i+1} to the DE. Thus, "consistency of the method" (meaning that the **sampled slope method $\phi(t_i, y_i; h)$ is equivalent to the original DE**) is verified by showing that the limit of $\lim_{h \to 0} \tau[h]$ = $\lim_{h \to 0} \{ (y_{i+1} - y_i)/h - \phi(t_i, y_i; h)\}$ =0. This follows because $\lim_{h \to 0} (y_{i+1} - y_i)/h \to y'$, and the $\lim_{h \to 0} \phi(t_i, y_i; h) = f(t_i, y_i)$, so that in this limit the "method" leads back to the original DE $y' = f(t_i, y_i)$.

Special Numerical Considerations for IVP Techniques

5.1.1 Examples of Consistency Test

Examples of Consistency Test

Single Step Method

Euler: $w_{i+1} = w_i + h\underbrace{f(t_i, w_i)}_{\equiv \phi_{Euler}}$ $\phi_{Euler}(t_i, w_i; h) \equiv f(t_i, w_i)$

obvious since $f(t_i, y_i)$ indep of h
$$\therefore \lim_{h \to 0} \phi_{Euler}(t_i, y_i; h) \equiv f(t_i, y_i)$$

Taylor y_{i+1} abt t_i: $y_{i+1} = y_i + y_i'h + y_i''\dfrac{h^2}{2} + O(h^3)$

Test: $\lim_{h \to 0} \left| \dfrac{y_{i+1} - y_i}{h} - \phi_{Euler}(t_i, y_i; h) \right| = \left| \dfrac{y_i'h + y_i''\frac{h^2}{2} + O(h^3)}{h} - f(t_i, y_i) \right| = \lim_{h \to 0} \left| \underbrace{y_i' - f(t_i, y_i)}_{=0 \text{ (Diff l Eqn)}} + y_i''\dfrac{h}{2} + O(h^2) \right| = 0$

Two-Step Method

Modified Mdpt: $w_{i+1} = w_{i-1} + 2hf(t_i, w_i)$ Difference Eqn with w_i

$y_{i+1} = y_{i-1} + 2hf(t_i, y_i)$ Difference Eqn with y_i

Taylor abt t_i:
$y_{i+1} = y_i + y_i'(h) + y_i''\dfrac{(h)^2}{2} + O(h^3)$ $y_{i-1} = y_i + y_i'(-h) + y_i''\dfrac{(-h)^2}{2} + O(h^3)$

Test: $\cancel{y_i} + y_i'(h) + \cancel{y_i''\dfrac{(h)^2}{2}} + O(h^3) = \cancel{y_i} + y_i'(-h) + \cancel{y_i''\dfrac{(-h)^2}{2}} + O(h^3) + 2hf(t_i, y_i)$

$\lim_{h \to 0} 2y_i'(h) = 2hf(t_i, y_i) + O(h^3)$ **Difference Eqn Yields Differential Eqn in limit as $h \to 0$**

yields $y_i' = f(t_i, y_i)$

Here are two examples of the Consistency Test.
In the linear Euler method $\phi^{(1)}(t_i, w_i; h)$ is linear and upon replacement of w_i by y_i we form the expression $|(y_{i+1} - y_i)/h - \phi^{(1)}(t_i, y_i; h)|$ and show that it approaches zero by
(i) making a Taylor expansion of y_{i+1} about t_i to 2nd order in h, and
(ii) replacing $\phi^{(1)}(t_i, y_i; h)$ by its Euler value $f(t_i, y_i)$ to find the bracketed term $y_i' - f(t_i, y_i)$ which is identically zero by virtue of the differential equation plus a term $y_i''*h/2$ which vanishes linearly.
Thus consistency of the Euler Method is established. This case is obvious since $\phi^{(1)}(t_i, y_i; h) = f(t_i, y_i)$ is independent of h so the lim $t \to 0$ is trivial.
In the Modified Midpoint Method (2-step) the slide established consistency by a simple substitution of the two Taylor series for y_{i+1} and y_{i-1} to establish that the differential equation is satisfied. Alternately we may re-write the midpoint method as follows
$$w_{i+1} = w_{i-1} + 2\,h\,f(t_i, w_i) = (w_i - w_i) + w_{i-1} + 2\,h\,f(t_i, w_i)$$
$$= w_i + h\,\{-(w_i - w_{i-1})/h + 2\,h\,f(t_i, w_i)\} = w_i + h\,\phi^{(2)}(t_i, w_i; h)$$
where we have added and subtracted w_i in the second equality and set that result equal to the standard form from which we identify $\phi^{(2)}(t_i, w_i; h) = \{(w_{i-1} - w_i)/h + 2\,f(t_i, w_i)\}$ for the Modified Midpoint method. Thus forming the expression
$$|(y_{i+1} - y_i)/h - \phi^{(2)}(t_i, y_i; h)| = |(y_{i+1} - y_i)/h - \{(y_{i-1} - y_i)/h + 2\,f(t_i, y_i)\}|$$
and now expanding all terms to 3rd order and again using the DE and canceling like terms we are left with a dominant term of order h which vanishes in the limit and hence establishes consistency of the Modified Midpoint Method.
Note that when a method $\phi^{(n)}(t_i, w_i; h)$ is derived from a nth order Taylor expansion, it is necessarily consistent; the consistency test is useful only in situations for which we are given an **non-vetted method** and wish to make certain that it is consistent before we expend any effort in programming it up.

Special Numerical Considerations for IVP Techniques

5.1.2 Consistency, Convergence, Stability Theorem: Single Step Case

Theorems Relating Stability to Consistency & Convergence-1

1. Single Step Methods:

- Theorem: Given IVP $\quad y' \equiv \dfrac{dy}{dt} = f(t, y(t)) \quad t \in [a,b] \quad IC: y(t=a) = \alpha$

  & Single step method $\quad w_0 = \alpha \quad w_{i+1} = w_i + \underbrace{h\phi(t_i, w_i; h)}_{Method}$

- **If**
  i. Continuous $\quad \phi(t_i, w_i; h) \in C$
  ii. Lipschitz on $\phi \quad |\phi(t_i, w; h) - \phi(t_i, \overline{w}; h)| \leq L|w - \overline{w}| \; ; \; h \in [0, h_0] \; \& \; D: \{t \in [a,b]; w \in (-\infty, \infty)\}$

- **Then**
  i. Numerical Difference Method is *Stable*
  ii. Difference Method Soln is *Convergent* iff Eqn *Consistent*

  $$\lim_{h \to 0} |y_{i+1} - w_{i+1}| = 0 \quad \Longleftrightarrow \quad \lim_{h \to 0} \tau_{i+1}^{trunc}[h] = 0$$

  $$\Rightarrow \lim_{h \to 0} \left| \frac{y_{i+1} - y_i}{h} - \phi(t_i, y_i; h) \right| = y' - \phi(t_i, y_i; 0) = 0 \quad \Rightarrow y' = f(t_i, y_i) = \phi(t_i, y_i; 0) \quad \left\{ \begin{array}{l} \text{For } h=0 \\ \text{reduces to} \\ \text{Euler Method} \end{array} \right.$$

  iii. Local/Global Trunc. Error Relationship

Local Error $\quad \varepsilon_{i+1}(h) = |y_{i+1} - y_i - h\phi| \; ; \; e(h) = \max\{\varepsilon_i(h)\}$

Global Error $\quad E_{i+1}(h) = |y_{i+1} - w_{i+1}| \leq \left(\dfrac{e^{L(t_{i+1}-a)}}{L} \right) \cdot \dfrac{e(h)}{h} = \dfrac{\tau(h)}{L} e^{L(t_{i+1}-a)}$

Global Error is Down one order in "h"

The relation between Stability, Consistency, and Convergence is quite different for Single Step Methods than it is for Multi Step Methods which have *parasitic solutions* that arise in the solution of the multistep difference equation. We need to develop some analysis tools in order to fully understand this difference; however, for single step methods the theorem is quite simple and states:

If the method $\phi(t_i, w; h)$ is ***continuous and satisfies the Lipschitz condition*** in the domain D and also for stepsize values in the range $h \in [0, h_0]$ then
Stability: the method is stable,
Consistency & Convergence: the DfE is consistent with the DE *if and only if* the estimate w_i converges to the true solution y_i (*i.e.*, consistency and convergence are equivalent to one another), and
Cumulative Error: Global truncation error is down one order in "h" relative to the local truncation error (as derived for the Euler method). Thus if the local truncation error for an nth order method $\phi^{(n)}(t_i, w; h)$ is $\varepsilon_i = kh^{n+1}$ then the global error accumulated at time t_{i+1} depends upon "h" as

$$E_{i+1}[h] = \tau[h]/h = kh^{n+1}/h = kh^n.$$

Thus, for single step methods, stability is guaranteed whenever the "method" given by the function $\phi^{(n)}(t_i, w; h)$ is continuous and satisfies a Lipschitz condition in "w".
Note 1: The requirement that the slope function $f(t, y(t))$ be ***continuous and satisfy the Lipschitz condition*** in the domain D insures that the IVP has a *unique solution* and that the problem is *well-posed*. The conditions stated here involve the same requirements for the **difference method $\phi(t_i, w; h)$ not the original slope function $f(t, y(t))$** in the statement of the IVP; moreover, the domain is also extended to include a range of stepsizes h between $[0, h_0]$.

Special Numerical Considerations for IVP Techniques

5.1.3 Analytic Solutions: Stability

Analytic Solutions: Stability

- **Standard of Stability**

$$y' = Ay \; ; \; y(0) = \alpha$$

$$y = \alpha e^{At}$$

Stable to small perturbations

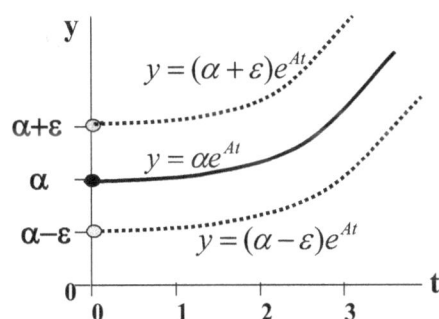

- **Standard of Instability**

$$y' = -y(1-y) \; ; \; y(0) = \alpha$$

$$y = \frac{\alpha}{\alpha + (1-\alpha)e^t}$$

Unstable for $\alpha = 1$
Solution changes radically
For small perturbations
about $\alpha = 1$

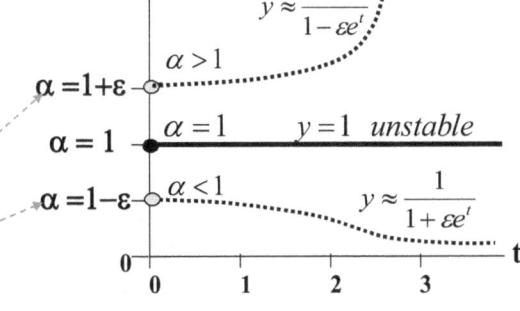

This slide is repeated for convenience to remind us of the **standard of "analytic" stability**. In the next slide, we compare results for the single step Euler method and the 2-step modified midpoint method applied to this analytically stable problem and explore the **numerical stability** of these two methods

We have seen in the slope plot that besides the issues of existence and uniqueness over a specified domain, the solution may become unstable and approach infinity for some specifications of the initial condition (IC). Even though this may be unavoidable over large changes in IC, we would like to insure that it does not happen when there is a **small perturbation** ε on an initial condition from say $\alpha = 1$ to $1+\varepsilon$ or $1-\varepsilon$. This property is extremely important because in numerical solutions there is always a small perturbation by virtue of the limited precision on conversion from floating point to machine numbers; that is, the analytic solution may be fine at a given $y(0) = 1$, but the small perturbation $y(0) = $ float(1)=$1+\varepsilon$ (or $1-\varepsilon$) caused by conversion of this IC to a machine represented number may yield a divergent solution. In this slide, the question of stability to small perturbations is addressed by considering two standard DEs with perturbed ICs $y(0) = 1+\varepsilon$ (or $1-\varepsilon$), which are taken as the standards of stability and instability. The DEs and their solutions are given in the top and bottom panels of the slide respectively. The top figure illustrates that the perturbed solutions "track" the analytic solution by remaining close to it for the stable exponential case. The bottom figure illustrates that for the instability standard, the perturbed solutions do not always "track" the analytic solution; more specifically, for the specific case $\alpha=1$, the analytic solution is simply a constant $y=1$, but the two perturbed solutions do not follow the analytic solution and in fact have completely different behaviors with $y(0) = 1+\varepsilon$ approaching infinity and $y(0) = 1-\varepsilon$ approaching zero asymptotically. Clearly the constant, $y=1$ solution is unstable. Note that the solution to the unstable DE is obtained by separating variables and making a partial fraction expansion to obtain $-dt = dy/[y(1-y)] = dy/y + dy/(1-y)$ which integrates to give

$$-t + \ln(C) = \ln(y) - \ln(1-y) = \ln(y/(1-y)) \rightarrow y/(1-y) = Ce^{-t} \text{ or } y = Ce^{-t}/[1+Ce^{-t}] = C/[C+e^t]$$

Applying the IC $y(0) = \alpha = C/[1+C]$ yields $C = \alpha/(1-\alpha)$ and the solution $y = \alpha / [\alpha + (1-\alpha) e^t]$

5.2 Multistep Methods & Parasitic Solutions - 1

Multistep Methods & Parasitic Solutions - 1

- **Analytic Soln** – Inherent Stability/Instability – Recall Standards
  - Inherent stability $\quad y' = Ay \,;\; y(0) = \alpha \,;\;$ Soln: $y = \alpha e^{At}$
  - Inherent instability $\quad y' = -y(1-y) \,;\; y(0) = \alpha \,;\;$ Soln: $y = \dfrac{\alpha}{\alpha + (1-\alpha)e^{t}}$
- **Numerical Soln** –
  - *Instability* can occur even if the analytic solution is inherently *stable*
  - Numerical multistep algorithms can create *parasitic solutions*
- **Example: Stability Std** $\quad y' = Ay \,;\; y(0) = \alpha \,;\;$ Soln: $y = \alpha e^{At}$
  - **Single Step Euler:** $w_{i+1} = w_i + hf(t_i, w_i) = w_i + hAw_i = (1 + hA)w_i$

  $$w_i = \alpha(1+hA)^i \to \alpha e^{Aih} + O(h^2) \to \alpha e^{At}$$

  Trial Soln $w_i = \lambda^i$
  $\Rightarrow \lambda^{i+1} = (1+hA)\lambda^i$
  $\Rightarrow \lambda = (1+hA)$

  - Track Exp. Growth $hA > 0$ requires $1 < 1 + hA < e^{Ah} = \{1 + hA + \tfrac{1}{2}(hA)^2 + \cdots\}$ — *Exp Growth*
  - Track Exp. Decay $hA < 0$ requires $|1+hA| < 1 \Rightarrow -2 < hA < 0$ — *Exp Decay & "Blip"*
  - Stepsize Constraint $\Rightarrow h < \dfrac{2}{|A|}$

- **Multistep Mod Midpt:** $w_{i+1} = w_{i-1} + 2hf(t_i, w_i) = w_{i-1} + 2hAw_i$

  - Trial Soln Yields: $w_i = \lambda^i \Rightarrow \lambda^{i+1} = \lambda^{i-1} + 2hA\lambda^i \Rightarrow \lambda_\pm = hA \pm \sqrt{1 + (hA)^2}$
  - Two Roots: $\lambda_+ = hA + 1 + \tfrac{1}{2}(hA)^2 \cong e^{hA} \,;\; \lambda_- = hA - 1 - \tfrac{1}{2}(hA)^2 \cong -e^{-hA}$
  - General Soln: $w_i = C_+(\lambda_+)^i + C_-(\lambda_-)^i = C_+(e^{hA})^i + C_-(-e^{-hA})^i$

The stable IVP has a solution αe^{At} and **numerical stability** requires that solution to the difference equation track this exponentially increasing (A > 0) or decreasing (A<0) solution for all values of the parameter A. Using this IVP, we now compare the stability of a single step Euler method with a modified midpoint method (multistep method) and show that the Euler method remains stable provided we bound the stepsize by the relation h < 2/|A|, while the multistep method yields a parasitic solution which does not track the known analytic exponential solution and is therefore unstable.

The standard method of solution to a difference equation is to substitute a trial solution of the form $w_i = \lambda^i$ into the difference equation and solve for the roots of the resulting polynomial. The general solution to the homogeneous DfE is then a linear combination of the root(s) raised to the ith power.

Single Step Euler yields a single root and hence a solution $w_i = \alpha(1+Ah)^i$ which approximates an exponential (to 1st order in h) $\alpha e^{Aih} = \alpha e^{Ati}$ For hA>0 (1+Ah) >1 and when raised to the power i the Euler solution properly tracks the increasing exponential. However, for A<0 (1+Ah)=(1-|A|h) and for (1-|A|h)^i to track the decreasing exponential as the exponent "i" increases without limit, the absolute value of this quantity must be less than unity, *i.e.*, |(1-|A|h)| < 1 This can be guaranteed for all values of A by requiring both 1-|A|h < 1 & |A|h-1 <1 which constrains the stepsize as 0 < h |A| <2. Thus we must impose a positive stepsize bound 2/|A|, *i.e.*,
$$0 < h < 2/|A|.$$

Modified Midpoint Method (2-step) yields a pair of roots and the general solution has a term of the form e^{+ihA} which tracks the exponential (increasing or decreasing) and a second "parasitic term" $(-1)^i e^{-ihA}$ which has the opposite exponential behavior e^{-ihA} and also oscillates in sign $(-1)^i$ with each time step. We consider this solution in more detail on the next slide.

Special Numerical Considerations for IVP Techniques

5.2.1 Multistep Methods & Parasitic Solutions-2

Multistep Methods & Parasitic Solutions-2

General Soln: $\quad w_i = C_+ e^{ihA} + C_-(-1)^i e^{-ihA}$

IC: $t_i = ih \quad i = 0: \quad w_0 = \alpha = C_+ e^{0hA} + C_-(-1)^0 e^{-0hA} \quad \Rightarrow \quad C_+ = \alpha - C_-$

Specific Soln: $\quad w_i = (\alpha - C_-)e^{ihA} + \underbrace{C_-(-1)^i e^{-ihA}}_{\text{parasitic term}} + O((hA)^3)$

Effects of RO Error - parasitic solns

No RO Error $\quad C_- = 0 \quad w_i = \alpha e^{ihA}$ $\qquad$ Well behaved – tracks exponential growth or decay

With RO Error $\quad C_- = \varepsilon \ll \alpha \quad w_i = (\alpha - \varepsilon)e^{ihA} + \underbrace{\varepsilon(-1)^i e^{-ihA}}_{\text{parasitic term}} \Rightarrow \begin{cases} hA > 0 & \varepsilon(-1)^i e^{-ihA} \to 0 \quad \text{Sign alternates \& Dies out} \\ hA < 0 & \varepsilon(-1)^i e^{+i|hA|} \to \pm\infty \quad \text{Sign Alternates \& Diverges} \end{cases}$

1) Parasitic Term does not track the correct solution no matter how small the stepsize h

2) Does not obey truncation error estimates and is unaffected by changes in h

3) Parasitic terms occur in all Multistep Methods

4) Euler solution has no parasitic term – the solution behaved correctly provided the stepsize was made small enough $\qquad h < \dfrac{2}{|A|}$

5) Important to know if **divergent parasitic** solutions are generated numerically!

The general solution of the difference equation now has two arbitrary constants (C_+ and C_-), but there is only one IC, namely for i=0 $w_0 = \alpha$, so we find $C_+ = \alpha - C_-$ where the coefficient of the parasitic term is unknown; this leads to a one parameter family of solutions with C_- undetermined. If, for example, we have $\alpha = 1/3$ then to 8 sd we set the coefficient of the "good solution" to $C_+ = .33333333$, and since $C_- + C_+ = \alpha$, the coefficient for the parasitic solution is the small "round off" difference $C_- = 1/3 - .33333333 = .333 \times 10^{-9}$ which results from the 8sd assignment of the initial value to C_+. Although, initially the solution will track the correct exponential, the initial small error grows exponentially with the time index "i" and eventually becomes the dominate term so the solution diverges. Thus, if we ignore RO error, we effectively set $C_- = 0$ and have $w_i = .33333333\, e^{ihA}$ which "tracks" both exponential growth and decay. On the other hand, including RO error the solution is

$$w_i = .33333333\, e^{ihA} + .333 \times 10^{-9} (-1)^i e^{-ihA}$$

For $hA > 0$ it accurately tracks the growing exponential since the 2nd parasitic term has the opposite exponential behavior and decays with time and eventually dies out, leaving the "good solution".
On the other hand, for $hA < 0$ the story is quite different because the decaying exponential in this case has a contrary parasitic term which **grows** and eventually dominates for large time values "ih".
These results are completely independent of the stepsize because we obviously do not change this *contrary behavior of the parasitic solution* simply by changing the stepsize. Moreover, truncation error estimates are useless when divergent parasitic solutions are present since they make RO error grow without bound and eventually swamp the correct solution. It is important to note that parasitic solutions never occur in single step methods because there is only one root; multistep methods always generate multiple roots and hence possibly destructive parasitic solutions. Clearly, it is important to be able to determine just when parasitic solutions are transitory and can be safely ignored. We have seen such a case above $hA > 0$ (exponential growth) in which the parasitic solution had contrary behavior and eventually decayed to zero leaving the correct solution. Even though single step solutions do not generate harmful parasitic solutions, they may still diverge if the stepsize is too large. In the Euler case we required $h < 2/|A|$ to assure stability. This defines a region in the h-A plane for stable solutions and in general it is important to determine this region the h-A plane for both single step and multistep methods.

Special Numerical Considerations for IVP Techniques

5.2.2 Characteristic Equation & Stability

Characteristic Equation & Stability

- **General Multistep Diff Eqn**
$$w_{i+1} = \underbrace{a_{m-1}w_i + a_{m-2}w_{i-1} \cdots + a_0 w_{i-(m-1)}}_{m\text{-point backward interpolation}} + \underbrace{hF(t_i, w_{i+1}, w_i, \cdots, w_{i-(m-1)}; h)}_{\text{Method - depends on "h"}}$$

  'm' Start Values $w_0 = \alpha \,;\, w_1 = \alpha_1 \,;\, w_2 = \alpha_2 \cdots\,;\, w_{m-1} = \alpha_{m-1}\,;\, \text{for } i = m-1, m-2, \cdots, N-1$

  $i = 3; m = 4: \quad w_4 = a_3 w_3 + a_2 w_2 + a_1 w_1 + a_0 w_0 + hF(t_3, w_4, w_3, w_2, w_1, w_0; h)$

- **Multistep Solution to Trivial Problem** $y'(t) = f(t,y) = 0 \Rightarrow F(t_i, w_i, w_{i-1}, \cdots, w_{i-m+1}; h) = 0$

  $y(t) = const. = \alpha$ **Assumption**

- **Trial Soln** $w_i = \lambda^i$ **yields**

  $w_{i+1} = a_1 w_i + \cdots + a_m w_{i-(m-1)}$
  $\lambda^{i+1} = a_1 \lambda^i + \cdots + a_m \lambda^{i-(m-1)}$
  $\lambda^{i-(m-1)}\{\lambda^m - a_1 \lambda^{m-1} - a_2 \lambda^{m-2} - \cdots - a_m\} = 0$

- **Characteristic Polynomial** $\boxed{\lambda^m - a_1 \lambda^{m-1} - a_2 \lambda^{m-2} - \cdots - a_m = 0}$

- $y(t) = const. = \alpha$ explicitly means: $w_{i+1} = w_i = w_{i-1} = \cdots = w_{i-m+1} = \alpha$
  Thus substitution of these values back into Diff Eqn yields the sum identity

  $\alpha = a_1 \alpha + a_2 \alpha + \cdots + a_m \alpha \Rightarrow \boxed{1 = a_1 + a_2 + \cdots + a_m}$ **Sum Identity**

- $\lambda = 1$ satisfies the Char. Poly since substitution yields the sum identity

The general form for multistep methods can be written as shown in the first equation; this equation expresses w_{i+1} as a linear combination or "weighted average" of "m" previous w_i's plus h times a linear combination of slopes evaluated at m previous w_i's and t_i s ($t_i = ih$). Solution of this general difference equation requires m-start values, only one of which, $w_0 = \alpha$ is given by the IC of the IVP; the remaining m-1 values must be generated by say an RK4 single step method. (Simple example is the modified midpoint method on Slide#5-3) Consider now a very trivial IVP $y'(t)=0$; $w_0 = \alpha$ which has the solution $y=const. = \alpha$; note in this case the slope function $f(t,y)$ is zero so the sum of slopes part of the general form $h \phi(t_i, w_i, ... w_{i-m+1}; h) = 0$. Now, it might seem that such a trivial problem with a constant solution should have no stability issues - not so!! We are still are left with an **m-step difference equation** $w_{i+1} = a_1 w_i + a_2 w_{i-1} + ... + a_m w_{i-m+1}$ to solve for a constant. Substituting the trial solution $w_i = \lambda^i$ into the difference equation leads to an mth degree polynomial in λ known as the **characteristic polynomial** and displayed in the boxed equation. The general solution is therefore a linear combination of each root raised to the ith power and this leads to possibly destructive parasitic solutions. Since the solution to this trivial DE is a constant we have $w_i = \alpha$ for all i, and substitution into the difference equation leads to the boxed sum identity stating that the "a" coefficients sum to unity. Moreover, we note that $\lambda = 1$ is easily verified to always be one solution of the characteristic polynomial since substitution results in the "sum identity for the "a" coefficients.

Nominally defining $\lambda_1 = 1$ to be first root of the polynomial we see that the "good" part of the general solution is in the term $a_1(\lambda_1)^i = a_1(1)^i = a_1$; hence if we set $w_0 = a_1 = float(\alpha)$ with a given precision (say 8 significant digits), then the remaining terms in the general solution must sum to the small difference $\Delta\alpha = \alpha - float(\alpha)$. Again the key is to ensure that these remaining terms do not "grow" and eventually swamp the "good" constant solution term. The next slide describes the required properties of the polynomial roots for this to be the case.

5.2.3 Root Structure, Parasitic Solutions, Stability

Root Structure, Parasitic Solutions, Stability

- **Parasitic Solns:**
$$m - \text{roots}: \lambda_1 = 1, \lambda_2, \lambda_3, \cdots, \lambda_m$$
$$w_i = \underbrace{\alpha \cdot (1)^i}_{\text{Exact Soln}} + \underbrace{c_2 \cdot (\lambda_2)^i + c_3 \cdot (\lambda_3)^i \cdots + c_m \cdot (\lambda_m)^i}_{\text{Parasitic Solns -Coeff Generated by RO Error}}$$

- Roots **Outside** Unit Circle yields Exponential Growth: λ^i for $|\lambda| > 1$
- Multiple Roots **On** Unit Circle yields Linear Growth: $(c_2 + c_3 \cdot t) \cdot (\lambda_2)^i$ for $\lambda_2 = \lambda_3 = 1$

- **Root Condition** $|\lambda| \leq 1$ & only *simple* roots on unit circle $|\lambda| = 1$
  no double roots, e.g., $\lambda = i, i$ or $\lambda = 1,1$ but $\lambda = -1,1$ OK
- Stability Definitions
  - **Strongly Stable**: Root Cond. Satisfied & $\lambda = 1$ only root with $|\lambda| = 1$
  - **Weakly Stable**: Root Cond. Satisfied & more than one root with $|\lambda| = 1$ (e.g., 1, i)
  - **Unstable**: Root Cond. Not Satisfied

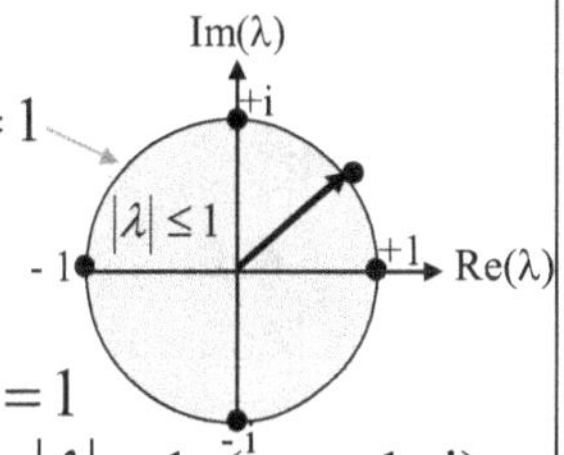

The boxed equation shows the general solution to the DfE as a linear combination of the m roots of the characteristic polynomial raised to the ith power. Note that the leading term for root $\lambda =1$ gives the "good solution" which approximates the exact solution. The remaining m-1 terms for polynomial roots $\lambda_2, \lambda_3, ..., \lambda_m$ are the parasitic terms generated by RO errors and these must not increase without limit.

In order to avoid exponential growth with increasing exponent "i" we require that all root magnitudes $|\lambda| \leq 1$; we already have the "good" root at $\lambda =1$, and we must require that all other roots be **on or within the unit circle in the complex plane** as shown by the shaded area in the figure. If all the roots are distinct, meaning no double, triple, ... , multiple roots, then we are finished. However, for example, if we have a double root $\lambda_2 = \lambda_3 = 1$, the solution to the difference equation yields a term of the form $(c_1+c_2 \, t_i)(\lambda_2)^i = (c_1+c_2 \, i*h)(1)^i$ which although not exponential increases linearly with index "i" and again eventually swamps the "good" solution. Thus, in order to avoid harmful parasitic solutions we must require the following

Root Condition: (i) All roots are on or within the unit circle $|\lambda| \leq 1$ and
 (ii) there are only simple roots on the unit circle ($|\lambda| = 1$)

(Note that this excludes double roots such as (i,i) and (1,1), etc. , but allows the two roots (1, i) , (i,-i), etc., which are not double roots). Thus, if the above **Root Condition is not satisfied**, then the solution is **unstable**; two "strengths" of stability are defined as **Strongly Stable** if there is only a **single root** with magnitude unity $|\lambda| =1$ (*i.e.* the "good" solution) and **Weakly Stable** if there is **more than one root** with magnitude unity (but still not multiple roots), *e.g.*, two roots (1, i) , (i,-i), *etc.*, which have $|\lambda| =1$ but are distinct.

Special Numerical Considerations for IVP Techniques

5.2.4 Consistency, Convergence, Stability Theorem: Multi-Step Case

Theorems Relating Stability to Consistency & Convergence-2

2. Multistep Methods

$$w_{i+1} = \underbrace{a_{m-1}w_i + a_{m-2}w_{i-1} \cdots + a_0 w_{i-(m-1)}}_{m\text{-point backward interpolation}} + \underbrace{hF(t_i, w_{i+1}, w_i, \cdots, w_{i-(m-1)}; h)}_{\text{Method - depends on "}h\text{"}}$$

$$w_0 = \alpha\;;\;w_1 = \alpha_1\;;\;w_2 = \alpha_2 \cdots;\;w_{m-1} = \alpha_{m-1}\;;\;\text{for}\;i = m-1, m-2, \cdots, N-1$$

- **Root Condition** $\Leftrightarrow$ **Stability** (Characteristic Polynomial)
- If *Consistent* with DE, **Then**: *Root Cond* (Stability) $\Leftrightarrow$ *Convergence*
- **Note 1:** *Consistency* as $h \to 0$ requires both
  i. Local Trunc Error $\to 0$ $\quad \lim_{h\to 0} \tau(h) = \lim_{h\to 0} \dfrac{\varepsilon_{trunc}}{h} = 0$
  ii. Computed Start Values $\to$ True Values $\quad \lim_{h\to 0} \alpha_k = y(t_{i-k+1})\;;\;k=1,\cdots,m$
- **Note 2:** For single step methods Consistency and Convergence were equivalent.
- **Note 3:** However, for Multi-step methods **Consistency does not imply convergence** because of parasitic terms. Need **Root Condition** !! (prevents parasitic divergences). Multi-step methods also require the Lifschitz condition hold for both $\phi(t,y;h)$ and $f(t,y)$.

Given the previous discussion, it is not surprising that **Multistep Methods** are **Stable** if and only if the **Root Condition** holds. Furthermore, provided that the difference **Method is Consistent** (with the DE, *i.e.*, $\tau(h)\to 0$), the **Solution Converges** ($|y_i - w_i| \to 0$) if and only if it is **Stable (Root Condition holds)**. That is, *Consistency* alone does not imply *Convergence* of the solution; the existence of *parasitic terms requires the Root Condition* to hold as well.

Contrast this with the **Single Step Methods** in which **Stability** follows directly from continuity and the Lipschitz condition (no root condition since $\lambda = 1$ is the only root) and the **Solution Converges** ($|y_i - w_i| \to 0$) if and only if the **Method is Consistent** ($\tau(h)\to 0$).

Note the further complication for the Consistency Multistep Methods is that the additional start values must generated by a consistent method and they must also converge to the correct y_i-values as the stepsize $h \to 0$. Contrast this with the Single Step (self-starting) Methods for which **solution stability** just requires the method function $\phi(t,y;h)$ be continuous and satisfy a Lifschitz condition; once stability is established, consistency (with the original DE) is equivalent to convergence (of the solution) so if it converges, it is consistent and *vice-versa*, if it is consistent, it converges (See Theorem on Slide#5-4).

Special Numerical Considerations for IVP Techniques

5.2.5 Stability Analysis Examples

Stability Analysis Examples

1. **Self Starting Methods** $w_{i+1} = w_i + h\phi(t_i, w_i)$ (*Euler, RK4, Midpt, Huen, etc.*)
   Characteristic Poly: $\lambda^{i+1} - \lambda^i = \lambda^i(\lambda - 1) \Rightarrow \lambda = 1$ has only **one root**
   → **Strongly Stable**

2. **Modified Midpoint** $w_{i+1} = w_{i-1} + 2h \cdot f(t_i, w_i)$ (*2-Step Method*)
   Charac. Poly: $\lambda^{i+1} - \lambda^{i-1} = \lambda^{i-1}(\lambda^2 - 1) \Rightarrow \lambda = \pm 1$ has roots 2 simple on unit circle
   → *Weakly Stable* (Midpoint method on Slides#5-6, 5-7 gave parasitic terms)

3. **Adams-Bashforth 4** $w_{i+1} = w_i + \dfrac{h}{24}[55f_i - 59f_{i-1} + 39f_{i-2} - 9f_{i-3}]$
   Characteristic Poly: $\lambda^{i+1} - \lambda^i = \lambda^i(\lambda - 1) \Rightarrow \lambda = 1$ has only **one root**
   → **Strongly Stable**

4. **Milne** $w_{i+1} = w_{i-3} + \dfrac{4h}{3}[2f_i - f_{i-1} + 2f_{i-2}]$
   Charac. Poly: $\lambda^{i+1} - \lambda^{i-3} = \lambda^{i-3}(\lambda^4 - 1) \Rightarrow \lambda = \pm 1, \pm i$ 4 simple roots on unit circle
   → *Weakly Stable*

The roots of the characteristic polynomial are used to determine the stability characteristics strong, weak, or unstable for a number of common methods. Note that the characteristic polynomial refers to solution of the trivial constant solution for which the h ϕ term is set equal to zero.

1) First **all self starting methods** yield a single root λ=1 and are therefore strongly stable.
2) The **modified midpoint method** is a 2-step method because it relates w_{i+1} to w_{i-1} and w_i as shown and the characteristic polynomial yields two distinct roots on the unit circle λ=+1 and λ=-1 and is thus only weakly stable.
3) The **Adams-Bashforth** method is a 4-step method, but its characteristic polynomial yields but a single root λ=1 and is therefore strongly stable.
4) The **Milne Method** is also a 4-step method, but because it relates w_{i+1} to w_{i-3} it characteristic polynomial has 4 distinct roots on the unit circle, namely λ=+1 , λ=-1 , λ=+i , λ=-i and is therefore only weakly stable.
Thus it is clear that the structure of the linear combination of terms w_i , w_{i-1} , w_{i-2} , w_{i-3} is the key to the stability of the characteristic polynomial for the trivial DE with solution w_i = constant and even for this simple case we already have issues with stability that are independent of stepsize h and the specific slope function. However, the Characteristic Polynomial is only part of the story, for when we consider a non-trivial DE such as the *standard of stability* IVP both the stepsize h and the slope function parameter A, have a profound effect upon how the solution tracks the growing or decaying exponential solution to the IVP. Hence the method and its region of applicability in the A-h plane must also be considered. Recall that we have already seen such an effect in the single step Euler method where we needed to restrict the stepsize h < 2/|A| in order to insure stability.

5.3 Stiff Equations

Stiff Equations

- Tracking from *fast transient* to a *steady state* regimes.
- Stable Std IVP: $y' = Ay$; $y(0) = \alpha$ $y = \alpha e^{At}$
- Homogeneous: $y' = 0$ Euler Method
  Char. Poly for method; stepsize does not enter
- Inhomogeneous: $y' = Ay$ Euler Method needed $h < \frac{2}{|A|}$
  stepsize constraint to "track" decaying exponential
- This defines a **Region of Absolute Stability** in the h-A plane
- For example for $A = -50$
  require very small stepsizes $h < \frac{2}{|-50|} = .04$

- Moreover, the faster it decays ($A = -500$), the smaller the stepsize ($h = .004$) needed to track the decay and the larger the number of steps needed to reach the desired steady state solution!! Thus:
  - *Integration time & effort increases*
  - *Round Off errors build up*
- Use Implicit Methods:
  - Backward Euler $w_{i+1} = w_i + hf(t_{i+1}, w_{i+1})$
  - Trapezoidal $w_{i+1} = w_i + \frac{h}{2}[f(t_i, w_i) + f(t_{i+1}, w_{i+1})]$

Systems of equations often display solutions with multiple time scales that are difficult to track in a numerical solution simply because the stepsize required to track one "time scale" of the solution may affect the tracking of other components of the solution. These so-called **stiff differential equations** naturally arise in mechanical vibration problems, but can also occur in a single DE such as the rapidly exponential decay to a long steady state constant solution illustrated in the top figure. This solution effectively has two time scales because it has a transient solution that dies out quickly as well as a steady state solution which remains constant after the initial decay.

This latter solution is represented by our standard of stability IVP $y' = Ay$; $y(0) = \alpha$ with solution $y = \alpha e^{At}$; we previously found that the Euler method put an upper bound $h < 2/|A|$ on the stepsize needed to track the exponential decay for $A<0$. This inequality describes a hyperbola $hA = $"-2" to the right of which the solution is stable as shown in the lower figure. The hashed region in the figure is known as the Region of Absolute Stability in the A-h plane. (Note that the stepsize h is usually taken as positive so we are only interested in the upper half plane.)

For $A=-50$, we find that the Euler Method requires a very small stepsize $h < 2/|-50| = .04$ in order to track the exponential decay; moreover for an even sharper decay $A=-500$, the stepsize decreases by a factor of "10" to $h = .004$ and this means that we need a large number of Euler steps before we reach the steady state "constant" solution. This of course has two detrimental effects, namely (i) increased computational load and (ii) build up of round off error by virtue of the large number of operations. For these reasons, the Euler solution leads to a poor solution of the Standard IVP and it is found that implicit methods involving w_{i+1} on both sides of the equation work much better. Two such methods, the Backward Euler and the Trapezoidal Method are written down at the bottom of the slide for reference.

Special Numerical Considerations for IVP Techniques

5.3.1 Regions of Absolute Stability-General Discussion

> # Regions of Absolute Stability-General Discussion
>
> - **Homogeneous:** $y' = 0$ Euler Method : Char. Poly for method; stepsize does not enter
> - **Single Step:** Now use stepsize h as a parameter in method
>
>   - Test Eqn: $y' = Ay$; $y(0) = \alpha$; Soln: $y = \alpha e^{At}$ $\qquad w_{i+1} = w_i + \underbrace{h\phi(t_i, w_i; h)}_{\text{Method}}$
>   - Euler: $w_{i+1} = w_i + hf(t_i, w_i) = w_i + hAw_i = (1+hA)w_i$
>   - Trial Soln: $w_i = \lambda^i$ yields polynomial Q: $\lambda^{i+1} = (1+hA)\lambda^i \Rightarrow \underbrace{\lambda^i[\lambda - (1+hA)]}_{\equiv Q(hA, \lambda)} = 0$
>
> **Multistep:** General Form
> $$w_{i+1} = \underbrace{a_{m-1}w_i + a_{m-2}w_{i-1} \cdots + a_0 w_{i-(m-1)}}_{m\text{-point backward interpolation}} + \underbrace{hF(t_i, w_{i+1}, w_i, \cdots, w_{i-(m-1)}; h)}_{\substack{=h\int y'(t)dt = h\int Ay(t)dt \\ \cong hA(b_0 w_{i+1} + \cdots + b_m w_{i-m+1}) \;\leftarrow\; \text{Linear comb of m slopes "Aw}_k\text{"}}}$$
>
>   - Trial Soln: $w_i = \lambda^i$ yields polynomial Q:
>
> $$\boxed{Q(hA; \lambda) \equiv (1 - hAb_m)\lambda^m - (a_{m-1} + hAb_{m-1})\lambda^{m-1} - \cdots - (a_0 + hAb_0) = 0}$$
>
>   - m roots of Polynomial in $Q(hA; \lambda)$ Determine Region of Absolute Stability. AB4 yields:
> $$w_4 = w_3 + (h/24) \cdot [55f_3 - 59f_2 + 37f_1 - 9f_0] \Rightarrow w_4 = w_3(1 + (55hA/24)) + (h/24) \cdot [-59Aw_2 + 37Aw_1 - 9Aw_0]$$
> $$w_i = \lambda^i \text{ yields}: \quad \lambda^4 - (1 + (55hA/24))\lambda^3 + (59hA/24)\lambda^2 - (37hA/24)\lambda^1 + (9hA/24) = 0$$

The bound on the stepsize $h < 2/|A|$ for the Euler Method defined a Region of Absolute Stability in the A-h plane and this idea can be naturally extended to general Multistep Methods. As an exemplar for the multi-step derivation, we formally derive the polynomial $Q(hA; \lambda) = \lambda - (1+hA)$ for the Euler Method. This polynomial has a single root $\lambda = (1+hA)$ and in order to make the solution $(1+hA)^i$ decrease with exponent "i" (for the exponential decay case $A < 0$) we must have $|1+hA| < 1$ which leads to the stepsize inequality $h < 2/|A|$.

Similarly, the general Multistep Difference Equation (DfE) contains the leading linear sum of terms $(a_1 w_i + \ldots + a_m w_{i-m+1})$ that are independent of both A and h and the integral of $y' = f(t,y) = Ay$ which is expressed by the "sample slopes method" as a sum of slopes hA ($b_0 w_{i+1} + \ldots + b_m w_{i-m+1}$). Substituting $w_i = \lambda^i$ into this "explicit form" of the difference equation yields the polynomial $Q(hA; \lambda)$ involving "a"s, "b"s, h, and A given in the boxed equation. Again the first root must be $\lambda_1 = 1$ and the requirements that the each of the remaining (m-1)-roots of this polynomial have absolute value $|\lambda_k| < 1$ for $k = 2, 3, \ldots, m$ will each impose a set of constraints on h and A that must all be satisfied and hence map out a Region of Absolute Stability in the A-h plane.

Clearly, finding the roots of the polynomial, then finding the constraints, and finally mapping the region in the A-h plane is a difficult task except in certain simple situations. But the region is there and can be found using a computer algorithm. It is important to know the restrictions on the stepsize h which assures us that the parasitic solutions behave properly.

Special Numerical Considerations for IVP Techniques

5.3.2 Stiff Equations - Example

Stiff Equations - Example

Stiff IVP:	$y' = -50y + 100$; $y(0) = \alpha$ $\Rightarrow$	**Exact Soln:**	$y(t) = \underbrace{(\alpha - 2)e^{-50t}}_{\text{transient}} + \underbrace{2}_{\text{stdy state}}$

	Euler (Explicit)	**Bkwd Euler (Implicit)**
Numerical Setup:	$w_{i+1} = w_i + h(-50w_i + 100)$ $w_{i+1} = (1-50h)w_i + 100h$	$w_{i+1} = w_i + h(-50w_{i+1} + 100)$ $\Rightarrow$ $w_{i+1}(1+50h) = w_i + 100h$ $w_{i+1} = \dfrac{w_i}{(1+50h)} + \dfrac{100h}{(1+50h)}$
Analytical Compl Soln:	$w_{i+1} = (1-50h)w_i$ $w_i = \lambda^i \Rightarrow \lambda^{i+1} - (1-50h)\lambda^i = 0$ $\Rightarrow \lambda = (1-50h)$ $w_i = c\lambda^i = c(1-50h)^i$	$w_{i+1} = \dfrac{w_i}{(1+50h)}$ $w_i = \lambda^i \Rightarrow \lambda^{i+1} = \dfrac{\lambda^i}{(1+50h)}$ $\Rightarrow \lambda = \dfrac{1}{(1+50h)}$ $w_i = c\lambda^i = c\left(\dfrac{1}{1+50h}\right)^i$
Particular Soln:	$w_{i+1} = w_i = const. = b$ $b = (1-50h)b + 100h$ $b(1-1+50h) = 100h \Rightarrow b = 2$	$w_{i+1} = w_i = const. = b$ $\Rightarrow b = \left(\dfrac{b}{1+50h}\right) + \left(\dfrac{100h}{1+50h}\right)$ $\Rightarrow b\left(1 - \dfrac{1}{1+50h}\right) = \left(\dfrac{100h}{1+50h}\right)$ $\Rightarrow b = \left(\dfrac{100h}{1+50h}\right) \cdot \dfrac{1+50h}{50h} = 2$
General Soln:	$w_i = c(1-50h)^i + 2$	$w_i = c\left(\dfrac{1}{1+50h}\right)^i + 2$
IC $i=0$:	$w_0 = \alpha = c(1-50h)^0 + 2 = c + 2 \Rightarrow c = \alpha - 2$	$i = 0$: $w_0 = \alpha = c\left(\dfrac{1}{1+50h}\right)^0 + 2 = c + 2 \Rightarrow c = \alpha - 2$
Problem Soln:	$w_i = \underbrace{(\alpha-2) \cdot (1-50h)^i}_{\text{transient}} + \underbrace{2}_{\text{stdy state}}$	$w_i = \underbrace{(\alpha - 2) \cdot \left(\dfrac{1}{1+50h}\right)^i}_{\text{transient}} + \underbrace{2}_{\text{stdy state}}$
Stepsize Constraint:	$h < 2/50 = .04$	None !!

This slide solves a stiff IVP by setting up and actually solving the difference equations associated with the Euler and Backward Euler methods. This side-by-side comparison shows exactly why the implicit method yields a convergent solution without stepsize constraints. The IVP: $y' = -50y + 100$; $y(0) = \alpha$ has the analytic solution consisting of a transient decaying exponential and a steady state constant solution: $y(t) = (\alpha - 2)e^{-50t} + 2$.

The first row shows the numerical set up for each giving the two difference equations; the implicit Backward Euler equation has w_{i+1} on both sides, but is easily solved for w_{i+1} since $f(t,y)$ is linear in y.

In the second row the trial solution $w_i = \lambda^i$ is substituted into the homogeneous difference equation to yield the "complementary solution" consisting of a single root raised to the ith power and multiplied by an arbitrary constant C.

In the third row a particular solution is b=constant =2 is verified

In the fourth row the general solution is written down as the sum of the complementary and particular solutions

In the fifth row the initial condition for i=0, $w_0 = \alpha$ yields the constant C= α -2

In the sixth row the two solutions are written and we see that for Euler that $(1-50h)^i$ requires the stepsize constraint $h < 2/50$; but the implicit Backward Euler method converts that term to $1/(1+50h)^i$ and there is no stepsize constraint needed because the term in the denominator is always greater than 1 for any stepsize h and hence the term is always less than unity and decays exponentially with index "i" as required. Actual application of these two methods shows that the implicit method is far superior to the Euler Method.

5.3.3 Problem with Euler for Stiff IVP

Problem with Euler for Stiff IVP

- Take specific IC $y(0) = \alpha = 0$ Soln: $y(t) = \underbrace{-2e^{-50t}}_{\text{transient}} + \underbrace{2}_{\text{stdy state}}$

- At time $t = 0.1$, $y(t)$ is within 1% of its Steady State (SS) value $= 2$
- Thus for $t > 0.1$, since it is so close to SS, we would want to use large steps h
- *However,* to ensure convergence to SS, Euler constrains stepsize: h<2/50=.04 so as to "track" the transient (decaying exponential).
- Hence, if a "fixed" stepsize is used, the value h=.04 is determined by a term that dies out after just a few steps of size h.
- Furthermore, the **more rapidly** the transient dies out, the **smaller** the required h
- Variable stepsize is very costly in multistep methods because of restart costs. ➔ *Use Implicit Methods instead*!!

The exact solution to the IVP of the last slide for a specific value of $\alpha = 0$ is written down as $y(t) = 2[1-e^{-50t}]$ which starts at the origin and rapidly rises to a value of y=2. The rise is so rapid that at time t=0.1, y(0.1) is within 1% of its steady state value of y=2 and we would like to use large steps in the numerical solution in order to reach the long term steady state solution.

However, the Euler method will not track the analytic solution unless we use a small step of h =.04 ; thus the stepsize is fixed to be very small by the transient term that virtually dies out after "three of these h=.04 steps" since 3(.04) = .12 . The problem is that if we do not continue with this small stepsize the transient will start to grow again and we will never reach the steady state solution. Moreover, the more rapidly the transient dies out, the smaller the required stepsize needed to track it. Recalling that stepsize changes with multi-step methods are very costly because of restarts, we see that they are not a good option.

The backward Euler and other implicit methods work best for Stiff DEs; in fact most programmable calculators have a "check box" for special handling of stiff DEs.

Special Numerical Considerations for IVP Techniques

5.4 Summary of IVP Solutions

Summary of IVP Solutions

Method	Self-Starting	Local Trunc. Error	Global Trunc. Error	$f(t,y)$ Eval/Step		Stepsize Control
				No Ctrl	w/control	
RK4	Yes	$O(h^5)$	$O(h^4)$	4	11	Costly h/2 step
RKF45	Yes	$O(h^5)$	$O(h^4)$	6	6	Easy
AP-MC4	No	$O(h^5)$	$O(h^4)$	5	5*	Moderate Restart

* Re-starts of multi-step method increase number of functional eval/step

- All above methods are **Strongly Stable**
- Some Guidelines:
  - *Real time*, f(t,y) Evals expensive so choose **Predictor-Corrector AP-MC4**
  - *Possible Singularity* in domain: need stepsize control : **RKF45, AP-MC4**
  - *Stiff IVP* then use Implicit Methods: **Bkwd Euler, Trapezoid, Implicit Runge-Kutta**
  - *Fixed Output Interval H* for tabulation or long time runs use **Extrapolation Methods**

The table summarizes several important IVP solution techniques according to whether they are self-starting, order of local and global truncation errors, the number of functional evaluations per step, and the ease with which stepsize control can be implemented. All these methods are strongly stable. The bullets below the table indicate special application areas for these methods and reasons for their choice in each case according to the parameters in the table.

In real time situations where there is no time for a large number of f(t,y) evaluations, the Predictor-Corrector methods are often used (Slide# 3-6 to 3-8). If there is the possibility of a singularity in the integration domain stepsize control near the singularity may be required and methods which do this easily are the RKF45 (Slide# 2-27 to 2-28) and to a lesser extent because of the cost of restarts, the predictor-corrector method. For stiff DEs the Backward Euler (Slide# 5-14), and the Implicit Runge-Kutta (next Slide# 5-17), are good choices. Finally for tabulating a function using fairly large output intervals H to cover a long run in time (say 0 to 100H), Extrapolation Methods (Gragg) are quite efficient and accurate (Slide# 3-9 to 3-11).

Other situations may call for different trade-offs and that is why numerical algorithms are more of an "art-form" than a science, being more like recipes than strict formulations guided by some overriding principles. For this reason, one very popular source book is appropriately titled "Numerical Recipes."

Special Numerical Considerations for IVP Techniques

5.4.1 Implicit Runge-Kutta Family

Implicit Runge-Kutta Family

RK2 Implicit Family (Patel: Num Anal)

$$w_{i+1} = w_i + a_1 k_1 + a_2 k_2$$
$$k_1 = hf(t_i + (\alpha + \beta)h, w_i + \alpha k_1 + \beta k_2)$$
$$k_2 = hf(t_i + (\gamma + \delta)h, w_i + \gamma k_1 + \delta k_2)$$

Apply Newton-Raphson

k1 & k2 occur on **both sides**

One Specific Member of Family $a_1 = a_2 = 1/2;\ \alpha = \delta = 1/4;\ \beta = 1/4 - \sqrt{3}/6;\ \gamma = 1/4 + \sqrt{3}/6$

Single Step Equation $w_{i+1} = w_i + \frac{1}{2}(k_1 + k_2)$

Implicit Coupled Eqns for Step Parameters k_1 & k_2

$$k_1 = hf(t_i + (\frac{1}{2} - \frac{\sqrt{3}}{6})h,\ w_i + \frac{1}{4}k_1 + (\frac{1}{2} - \frac{\sqrt{3}}{6})k_2)$$
$$k_2 = hf(t_i + (\frac{1}{2} + \frac{\sqrt{3}}{6})h,\ w_i + (\frac{1}{2} + \frac{\sqrt{3}}{6})k_1 + \frac{1}{4}k_2)$$

Requires 2^d Newton-Raphson solution to
$f_1(k_1, k_2)=0$
$f_2(k_1, k_2)=0$

Here is a family of Implicit RK2 algorithms found in the book Numerical Analysis by Patel [Ref. 7, p. 302]. A particular member of this family is given on the lower half of the slide. Note that the update equation involves two slope parameters k_1 and k_2 which are obtained from the implicit equations directly below it. A 2-dimensional Newton-Raphson procedure must be used to simultaneously solve the coupled pair of non-linear equations for k_1 and k_2.

In addition to the direct computational costs, the Newton-Raphson method requires some careful algorithmic design because it converges rapidly only when we have a good initial guess and we must develop a procedure that provides such guesses "on-the-fly."

6 Direct Solution of Linear Systems

Direct Solution of Linear Systems

6.1 Preliminaries on Linear Systems

---

n x n Linear Matrix Systems

1) Deterministic Solutions
2) Iterative Solutions

We consider fully determined Linear systems in which there are an equal number of equations and unknowns; thus we will deal exclusively with systems of equation can be expressed in terms of square n x n matrices. We shall first discuss methods for finding explicit deterministic solutions by algebraic manipulation of the system of equations and then we will show how iterative solutions can be used to advantage when the number of equations becomes large.

Direct Solution of Linear Systems

6.1.1 Deterministic Solutions

---

n x n Linear Systems
1) Deterministic Solutions

- Gaussian Elimination
- Matrix Decomposition Ax=b → (LU)x=b
- Yields Solution of two easier problems
    1. Lc=b   Forward solution
    2. Ux=c   Backward solution
- Efficiency – Counting MAD
- Special Types of Matricies
- Order of operations, RO, & Pivoting
- Determinants & Direct Matrix Inversion
- Other Matrix Decompositions: Crout, Choleski, LDU

Gaussian Elimination is a straightforward technique for solving a linear system **Ax** = **b**, where **b** is the n-dimensional data vector and **x** is the n-dimensional solution or state vector and **A** is an n x n measurement matrix . It turns out that this process is equivalent to decomposing the original matrix **A** into the product of a pair of lower **L** and upper **U** triangular matrices which allows us to re-cast the original matrix equation into a pair of matrix equations that can be used to solve a large number of such matrix equations $Ax_1 = b_1$, $Ax_2 = b_2$, ... $Ax_n = b_n$, in a very efficient manner. The efficiency of a solution algorithm is determined by counting the total number of multiplies, adds, and divisions (MAD) required to find the solution to the matrix equation(s) and for an NxN matrix **A** this number of operations is generally of order N^3. Special techniques for matrices with a simple structure (such as tri-diagonal) reduce the number of operations significantly. We find that the order of operations in the solution procedure can have a significant effect on the accuracy of the results because of numerical round off error and this leads to the idea of "pivoting" which involves the swapping of rows (or columns) of the matrix **A** in order to reduce RO errors. Although, we can always solve a system of equations using Gaussian Elimination, it is desirable to have direct methods of finding the determinant and inverse of the matrix **A**. These are discussed as special cases of Gaussian and Jordan Elimination. Finally, the Crout, Choleski, and **LDU** matrix decompositions are developed and provide new multi-step algorithms for solving linear systems for special types of matrix **A**.

6.1.2 Iterative Solutions

---

n x n Linear Systems
2) Iterative Solutions

- Tools of the Trade
  - Vector & Matrix Norms
  - Eigenvalues & eigenvectors
- Iterative Methods
  - Jacobi
  - Gauss-Seidel
  - Successive Over Relaxation (SOR)
  - Efficiency
  - Error Estimates
  - Condition Number & Iterative Refinement

As an alternative to direct solution of the linear system $A\mathbf{x} = \mathbf{b}$ is to make an initial guess $\mathbf{x}^{(0)}$ which in turn produces an output $\mathbf{b}^{(0)} = A\,\mathbf{x}^{(0)}$ different from the correct vector value $\mathbf{b}$. We discuss three different iterative methods for iteratively adjusting the components of this initial vector $\mathbf{x}^{(0)}$ until the solution converges according to some criterion. The computational efficiency and accuracy are discussed for the Jacobi, Gauss-Seidel, and Successive Over Relaxation methods of iterative solution. Finally, the important concept of condition number $K(\mathbf{A})$ of a matrix $\mathbf{A}$ is discussed and shown to be key to assessing the accuracy of the solution to the linear system. A very simple procedure known as iterative refinement is shown to be a good way of dealing with the accuracy problems associated with highly ill-conditioned matrices.

Direct Solution of Linear Systems

6.1.3 State Determination from Measurements

State Determination from Measurements

- **Deterministic Linear system** $\mathbf{Ax = b}$

$$\mathbf{A} = \begin{bmatrix} a_{11} & a_{12} & a_{13} \\ a_{21} & a_{22} & a_{23} \\ a_{31} & a_{32} & a_{33} \end{bmatrix} \quad \mathbf{x} = \begin{bmatrix} x_1 \\ x_2 \\ x_3 \end{bmatrix} \quad \mathbf{b} = \begin{bmatrix} b_1 \\ b_2 \\ b_3 \end{bmatrix}$$

  meas. matrix   state   meas

- **Formal solution** $\mathbf{x = A^{-1}b}$ $(\mathbf{A}^{-1})_{ij} = \dfrac{Cof(\mathbf{A})_{ji}(-1)^{i+j}}{\det \mathbf{A}}$

- **2^d Matrix:** $\mathbf{A} = \begin{bmatrix} a & b \\ c & d \end{bmatrix} \quad \mathbf{A}^{-1} = \dfrac{1}{|\mathbf{A}|} \begin{bmatrix} d & -b \\ -c & a \end{bmatrix} \; ; \; |\mathbf{A}| = ad - bc$

$$\mathbf{AA}^{-1} = \dfrac{1}{ad-bc} \cdot \begin{bmatrix} a & b \\ c & d \end{bmatrix} \begin{bmatrix} d & -b \\ -c & a \end{bmatrix} = \dfrac{1}{ad-bc} \cdot \begin{bmatrix} ad-bc & -ab+ba \\ cd-dc & -cb+da \end{bmatrix} = \begin{bmatrix} 1 & 0 \\ 0 & 1 \end{bmatrix}$$

Typically a linear system is expressed by the matrix equation $\mathbf{Ax = b}$, where $\mathbf{b}$ is a set of measurements, $\mathbf{x}$ is the set of state parameters and $\mathbf{A}$ is the measurement matrix. The solution of this matrix equation is interpreted to mean that the measurements $\mathbf{b}$ may be used to infer the state of a system $\mathbf{x}$ by formally inverting the matrix $\mathbf{A}$ and expressing $\mathbf{x = A^{-1} b}$. The definition of the of $(i,j)^{th}$ element of the inverse matrix $(\mathbf{A}^{-1})_{ij}$ is given in terms of the cofactor corresponding to the transposed indices (j,i), i.e., $Cof(\mathbf{A})_{ji}$ The $Cof(\mathbf{A})_{ji}$ is defined as the determinant of the minor submatrix obtained by removing the jth row and ith column of the original matrix A. The determinant of a matrix $\det\mathbf{A} = |\mathbf{A}|$ is obtained by multiplying the leading element in a row (or column) by its associated minor determinant and adding all such products with alternating signs. The process is easily performed for n=2, 3 but becomes tedious for dimension n greater than 3.

A very simple example is to take two measurements of the position "b" of a object moving in a straight line with constant speed. The position measurement "b" is given by the scalar equation $b = x_0 + v_0 t$. For two measurements at times t_1 and t_2 we simply write this equation down twice to form the matrix equation $\mathbf{b = Ax}$

$$\begin{matrix} b_1 = 1 \cdot x_0 + t_1 \cdot v_0 \\ b_2 = 1 \cdot x_0 + t_2 \cdot v_0 \end{matrix} \Rightarrow \begin{bmatrix} b_1 \\ b_2 \end{bmatrix} = \begin{bmatrix} 1 & t_1 \\ 1 & t_2 \end{bmatrix} \cdot \begin{bmatrix} x_0 \\ v_0 \end{bmatrix}$$

where the measurement vector $\mathbf{b} = [b_1, b_2]^T$, the state vector $\mathbf{x} = [x_0, v_0]^T$, and the measurement matrix is $\mathbf{A} = [[1, t_1], [1, t_2]]$. Take the specific example of two measurements $b_1 = 2$ at $t_1 = 1$ & $b_2 = 3$ at $t_2 = 3$; the measurement vector is $\mathbf{b} = [2, 3]^T$ and the measurement matrix $\mathbf{A} = [[1, 1], [1, 3]]$. Now on the bottom half of the slide we show the inverse of a two dimensional matrix is obtained by swapping the diagonal elements, changing the sign of the off-diagonal elements and dividing by the determinant and we easily solve for the state vector.

$$\mathbf{x = A^{-1} b} = \tfrac{1}{2} [[3, -1], [-1, 1]] * [2, 3]^T = [3/2, 1/2]^T \; ; \; i.e., \; x_0 = 3/2 \text{ and } v_0 = 1/2.$$

6.1.4 Round Off Problems Again!

Round Off Problems Again!

- 2x2 Linear system $\quad \mathbf{Ax} = \mathbf{b}$

$$\mathbf{A} = \begin{bmatrix} 1.0000 & 2.0003 \\ 1.0002 & 2.0000 \end{bmatrix} \quad \mathbf{b} = \begin{bmatrix} 3.0003 \\ 3.0002 \end{bmatrix} \quad \mathbf{x}_{exact} = \begin{bmatrix} 1.0000 \\ 1.0000 \end{bmatrix}$$

- Find $\mathbf{x} = \mathbf{A}^{-1}\mathbf{b}$

$$\mathbf{A}^{-1} = \frac{1}{-.70006 \times 10^{-3}} \begin{bmatrix} 2.0000 & -2.0003 \\ -1.0002 & 1.0000 \end{bmatrix} = \begin{bmatrix} -2856.8 & 2857.2 \\ 1428.7 & -1428.4 \end{bmatrix}$$

$$\hat{\mathbf{x}} = \begin{bmatrix} -2856.8 & 2857.2 \\ 1428.7 & -1428.4 \end{bmatrix} \cdot \begin{bmatrix} 3.0003 \\ 3.0002 \end{bmatrix} = \begin{bmatrix} .90000 \\ 1.0000 \end{bmatrix}$$

The exact solution to this linear system is $\mathbf{x} = [1, 1]^T$; however direct computation to 5 significant digits using the inverse of $\mathbf{A}$ is shown to yield x-cap = $[.90000, 1.0000]^T$. It is easy to see why this happens because the matrix multiplication leading to the 1st component of $\mathbf{x}$ results in the subtraction of two numbers that are nearly equal and the loss of 4 significant digits. The 2nd component of $\mathbf{x}$ also involves the subtraction of two numbers that are nearly equal, but fortuitously does not lose any significant digits.

Thus we have the issue of potential loss of significant digits and round off problems are with us again!

6.2 Deterministic Solutions

n x n Linear Systems
Deterministic Solutions

Gaussian Elimination
Efficiency – Counting MAD
Pivoting
Determinants & Direct Matrix Inversion
Other Matrix Decompositions

In this first section on n x n linear systems of equations we consider deterministic solutions to a system of *n linear equations* in *n unknowns* given by the matrix equation **Ax=b**, where **A** is an n x n matrix and both unknown solution **x** and the known data **b** are n x 1 column vectors. Both direct matrix inversion **x=A^{-1} b** and linear system solution of the fundamental matrix equation **Ax=b** are explored. The latter is realized by Gaussian elimination which is a straightforward elimination technique that transforms all elements below the diagonal to zeros thereby recasting the system into one with an upper triangular matrix **U** and a new data vector **c**. The resulting triangular structure of the new matrix equation **Ux = c** allows for simple sequential "backward substitution" for the components of the solution vector **x** in the order $x_n, x_{n-1}, …, x_1$.

Gaussian elimination requires ~ n^3 multiplication (M), addition (A) and division (D) operations and therefore becomes problematic for large linear systems with say n =100 or more because the round off error inherent in finite digit computations reduces the accuracy of the direct solution. Round off errors can be reduced by scaling and pivoting elements in the matrix and by using special techniques adapted to particular types of matrices. For example, there is a technique specific to a tridiagonal matrix which drastically reduces the total number of operations to ~n required to solve the system of equations. Positive definite matrices do not require pivoting and other matrices have properties for which special techniques can be employed.

Even though we rarely invert a matrix to solve a linear system, the inverse and determinant of a matrix are of interest in their own right as they are needed to compute the *condition number* and the *eigenvalues* of the matrix, both of which shed light on how difficult a given matrix equation is to solve. We also discuss matrix decompositions that yield efficient solution methods for special matrices.

Direct Solution of Linear Systems

6.2.1 Forward Substitution Example

Forward Substitution Example

- **Lower Triangular Matrix L:** $\quad \mathbf{L} \cdot \mathbf{c} = \mathbf{b}$

$(E1): \quad (-2)c_1 + 0 + 0 = 8 \qquad c_1 = 8/(-2) = -4$

$(E2): \quad 1c_1 + (3)c_2 + 0 = 11 \qquad c_2 = [11 - (1)\cdot(-4)]/(3) = 5$

$(E3): \quad -2c_1 - 1c_2 + (6)c_3 = 9 \qquad c_3 = [\underbrace{9}_{b_3} - \{\underbrace{(-2)}_{l_{31}}\cdot\underbrace{(-4)}_{c_1} + \underbrace{(-1)}_{l_{32}}\cdot\underbrace{(5)}_{c_2}\}]/\underbrace{(6)}_{l_{33}} = 1$

- **Augmented Matrix** $\quad N=3, k=3$

$$\begin{bmatrix} -2 & 0 & 0 & \vdots & 8 & \vdots & -4 \\ 1 & 3 & 0 & \vdots & 11 & \vdots & 5 \\ -2 & -1 & 6 & \vdots & 9 & \vdots & 1 \end{bmatrix}$$
$\qquad \mathbf{L} \qquad\quad \mathbf{b} \quad \mathbf{c}$

$\qquad 8/(-2) \qquad\qquad step\#1$

$[11 - \{(1)\cdot(-4)\}]/(3) \qquad step\#2$

$[9 - \{(-2)\cdot(-4) + (-1)\cdot(5)\}]/(6) \qquad step\#3$

- **Generalize** $\quad c_k = \dfrac{1}{l_{kk}}\left\{b_k - \begin{bmatrix} l_{k1} & l_{k2} & l_{k3} & \cdots & l_{k,(k-1)} \end{bmatrix} \cdot \begin{bmatrix} c_1 \\ c_2 \\ c_3 \\ \vdots \\ c_{k-1} \end{bmatrix}\right\} = \dfrac{1}{l_{kk}}\cdot\left[b_k - \sum_{\alpha=1}^{k-1} l_{k\alpha} c_\alpha\right]$

The system of three equations can be written in the form $\mathbf{Lc} = \mathbf{b}$ where $\mathbf{L}$ is a lower triangular matrix with all non-zero elements on or below the main diagonal as shown. The solution to the system of three equations {E1, E2, E3} starts at the top solving the 1st equation for c_1: dividing b_1 by the single diagonal coefficient (-2) on the LHS; next the 2nd equation is solved for c_2: subtracting the (now known) 1st term $1c_1$ on the LHS from b_2 and dividing by 3 the (diagonal) coefficient of c_2; finally, the 3rd equation is solved for c_3: subtracting the (now known) terms $-2c_1$ -1 c_2 on the LHS from b_3 and dividing by 6 the (diagonal) coefficient of c_3. This process is called "forward substitution" because it solves the index "i" in increasing order.

Under the c_3 computation we have written out the terms in terms of the elements of the $\mathbf{L}$ matrix as follows:

$$c_3 = [\, b_3 - \{l_{31}c_1 + l_{32}c_2 \}\,] / l_{33}$$

This generalizes to

$$c_k = [\, b_k - \{l_{k1}c_1 + l_{k2}c_2 + ... + l_{k,(k-1)}c_{k-1} \}\,] / l_{kk}$$

and this expression can be written either in terms of an inner product of two vectors or equivalently the indexed sum shown on the slide. Note that we have set up an augmented matrix denoted by the partitioned matrix [$\mathbf{L}$:$\mathbf{b}$] in anticipation of further developments in which we will process rows of the original matrix $\mathbf{L}$ and the data vector $\mathbf{b}$ simultaneously to obtain the solution vector $\mathbf{c}$.

6.2.2 Backward Substitution Example

Backward Substitution Example

- **Upper Triangular Matrix U:** $\quad \mathbf{U} \cdot \mathbf{x} = \mathbf{c}$

$(E1):\ (2)x_1 - 5x_2 - 8x_3 = -3$
$(E2):\ 0 + (-1)x_2 + 3x_3 = -6$
$(E3):\ 0 + 0 + (-4)x_3 = 4$

$x_1 = [-3 - \{(-5) \cdot (3) + (-8) \cdot (-1)\}]/(2) = 2$
$x_2 = [-6 - (3) \cdot (-1)]/(-1) = 3$
$x_3 = 4/(-4) = -1$

$N=3,\ k=1$

- **Augmented Matrix**

$$\begin{bmatrix} 2 & -5 & -8 & \vdots & -3 & \vdots & 2 \\ 0 & -1 & 3 & \vdots & -6 & \vdots & 3 \\ 0 & 0 & -4 & \vdots & 4 & \vdots & -1 \end{bmatrix} \quad \begin{array}{l} [-3 - \{(-5) \cdot (3) + (-8) \cdot (-1)\}]/(2) \quad \text{step \#3} \\ [-6 - \{(3) \cdot (-1)\}]/(-1) \quad \text{step \#2} \\ 4/(-4) \quad \text{step \#1} \end{array}$$

$\quad\quad U \quad\quad c \quad x$

- **Generalize**

$$x_k = \frac{1}{u_{kk}}\left\{c_k - [u_{k,(k+1)}\ u_{k,k+2}\ u_{k,k+3}\ \cdots\ u_{k,N}] \cdot \begin{bmatrix} x_{k+1} \\ x_{k+2} \\ x_{k+3} \\ \vdots \\ x_N \end{bmatrix}\right\} = \frac{1}{u_{kk}}\left[c_k - \sum_{\alpha=k+1}^{N} u_{k\alpha} x_\alpha\right]$$

The system of three equations can be written in the form $\mathbf{Ux = c}$ where $\mathbf{U}$ is a upper triangular matrix with all non-zero elements on or above the main diagonal as shown. The solution to the system of three equations {E1, E2, E3} starts at the bottom solving the 3rd equation for x_3: dividing c_3 by the single diagonal coefficient (-4) on the LHS; next the 2nd equation is solved for x_2: subtracting the (now known) 3rd term $3x_3$ on the LHS from c_2 and dividing by -1 the (diagonal) coefficient of x_2; finally, the 3rd equation is solved for x_3: subtracting the (now known) terms $-5x_2 -8x_3$ on the LHS from c_1 and dividing by 2 the (diagonal) coefficient of x_1. This process is called "backward substitution" because it solves the index "i" from the bottom to the top in decreasing order.
Above the x_1 computation we have written out the terms in terms of the elements of the $\mathbf{U}$ matrix as follows:

$$x_1 = [\ c_1 - \{u_{12}x_2 + u_{13}x_3\ \}\ ] / u_{33}$$

This generalizes to

$$x_k = [\ c_k - \{u_{k,k+1}x_{k+1} + u_{k,k+2}x_{k+2} + ... + u_{k,N}x_N\ \}\ ] / u_{kk}$$

and this expression can be written either in terms of an inner product of two vectors or equivalently the indexed sum shown on the slide.
Note here again we have set up an augmented matrix denoted by the partitioned matrix [$\mathbf{U:c}$] in anticipation of further developments in which we will process rows of the original matrix $\mathbf{U}$ and the data vector $\mathbf{c}$ simultaneously to obtain the solution vector $\mathbf{x}$.

6.2.3 Gaussian Elimination Rationale

Gaussian Elimination Rationale

- Direct inverses
  - Solution is tedious for n>3
  - May be nearly singular detA~ 0 RO & cancellations
- Instead solve the system of equations for **x**
  - Direct Gaussian Elimination
  - Zero columns below diagonal
  - Upper Triangular Form
  - Adjoin and process together [**A**:**b**]
  - Sequence [**A**:**b**] → [**A**$_1$:**b**$_1$] ...→ [**U**:**b**$_{n-1}$]
  - Backward substitution

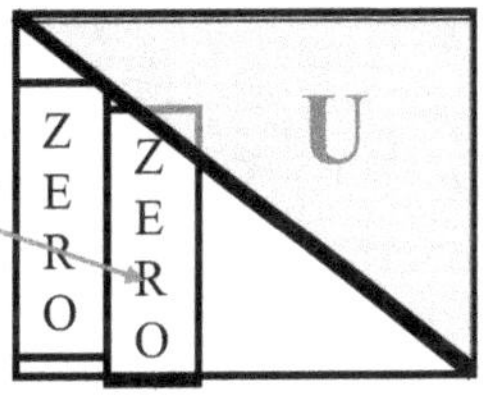

Using the formula for the inverse is tedious for dimension n>3 and, moreover, when the matrix has a determinant that is nearly zero (singular) numerical RO errors and cancellations become an issue. The subtraction of nearly equal numbers and subsequent division by a small number makes direct use of the formula problematic. As an alternative to finding the solution directly with the inverse $\mathbf{x} = \mathbf{A}^{-1}\mathbf{b}$, we may instead solve the system of equations **Ax**=**b** for **x**.

The Gaussian Elimination scheme reduces the original matrix **A** to an upper diagonal matrix **U** which then allows solution for the state **x** by the process of backward substitution as we discussed above for the matrix equation **U x** = **c**.

We define the "augmented matrix" for the problem **Ax**=**b** to be the matrix obtained by adding the n x1 column vector **b** to the original n x n matrix **A** to form an n x (n+1) matrix denoted by the "partitioned matrix" form [**A**:**b**]. Gaussian elimination uses row multipliers to systematically "zero out" the elements in each column of **A** below the diagonal to eventually leave the upper triangular matrix **U** shown in the diagram. If we perform these same operations simultaneously on the components of **b** in the augmented matrix we generate a sequence of new augmented matrices as shown in the sequence

$$[\mathbf{A}:\mathbf{b}] \rightarrow [\mathbf{A}_1:\mathbf{b}_1] \rightarrow [\mathbf{A}_2:\mathbf{b}_2] .... \rightarrow [\mathbf{A}_{n-1}:\mathbf{b}_{n-1}] = [\mathbf{U}:\mathbf{c}]$$

where **U** is the desired upper triangular matrix and the new augmentation vector $\mathbf{b}_{n-1}$ is defined to be **c**. This augmented matrix may be processed to solve for the components of the solution vector **x** using the same "backward substitution" formulas we generated earlier.

Direct Solution of Linear Systems

6.2.4 Gaussian Elimination Algorithm

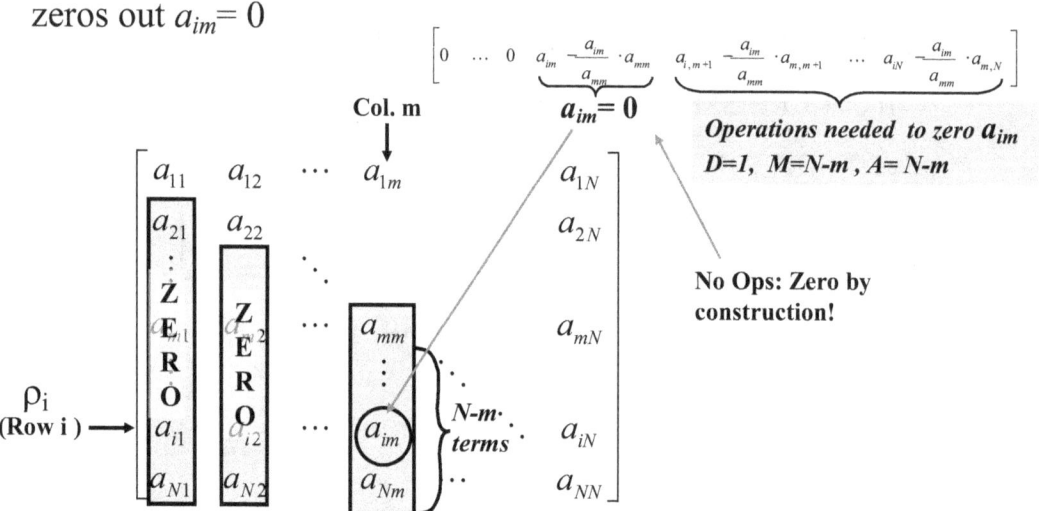

- Assume $m-1$ cols have already been zeroed out
- Consider row i in col m:

  Define row multiplier: $l_{im} = a_{im}/a_{mm}$

  Lin comb: $\rho_i \rightarrow \rho_i - l_{im}\,\rho_m = [0 \;\cdots\; 0 \; a_{im} \; a_{i,m+1} \;\cdots\; a_{iN}] - \dfrac{a_{im}}{a_{mm}} \cdot [0 \; 0 \;\cdots\; a_{mm} \; a_{m,m+1} \;\cdots\; a_{mN}]$

  zeros out $a_{im} = 0$

  $$\left[0 \;\cdots\; 0 \; a_{im} - \dfrac{a_{im}}{a_{mm}} \cdot a_{mm} \; a_{i,m+1} - \dfrac{a_{im}}{a_{mm}} \cdot a_{m,m+1} \;\cdots\; a_{iN} - \dfrac{a_{im}}{a_{mm}} \cdot a_{m,N} \right]$$

  $a_{im} = 0$

  Operations needed to zero a_{im}
  $D=1, \; M=N-m, \; A=N-m$

  No Ops: Zero by construction!

The Gaussian Elimination Algorithm reduces the original N x N matrix **A** to an upper diagonal matrix **U** by zeroing out the elements below the diagonal one column at a time as illustrated in the diagram. Working down a given column, each row element is individually zeroed out by replacing the (whole) target row with an appropriate linear combination of itself and the leading row. This is done for every row below the diagonal until only zeros reside below the diagonal in that column.

The process is illustrated at the stage where we have zeroed out the first (m-1) columns and we are working on the element a_{im}, having already zeroed out all elements above it except for the diagonal element a_{mm}. We can zero out this specific element a_{im}, by replacing the entire "row$_i$" by
" row$_i$ – (a_{im}/ a_{mm}) * row$_m$ "

This linear combination is written out explicitly on the slide and it should be noted that (i) this linear combination does indeed zero out the target element a_{im} and (ii) all elements to the left of a_{im} and to the left of a_{mm} are filled with zeros so they remain zero in the replacement row. Thus we do not undo any prior zeroing out.

An important aspect of any such procedure is the total number of multiplies, adds, and divides MAD needed to complete it. From the explicit representation on the slide we see that zeroing out this single target element a_{im} requires exactly M= N-m multiplies, A=N-m adds, and D=1 divisions. In making this count we do not count any operations involving already zeroed elements; nor do we count the implicit operations "designed" to zero out the target element, *i.e.*, " a_{im}– (a_{im}/ a_{mm}) * a_{mm}". In the next slide we use this counting result for the single element a_{im} to determine the total number of operations needed to complete the Gaussian Elimination transformation.

Direct Solution of Linear Systems

6.2.5 Counting Operations: M,A,D

Counting Operations: M,A,D

- m^{th} col: has $(N-m)$ elements below diagonal

$$D_m = 1 \cdot (N-m)$$
$$M_m = (N-m) \cdot (N-m)$$
$$A_m = (N-m) \cdot (N-m)$$

- Summing MAD ops over all but last column

  (which already has elements only above the diagonal)

$$D = \sum_{m=1}^{N-1} D_m = \sum_{m=1}^{N-1} 1 \cdot (N-m) = \frac{N \cdot (N-1)}{2}$$

$$M = \sum_{m=1}^{N-1} M_m = \sum_{m=1}^{N-1} (N-m) \cdot (N-m) = \frac{N \cdot (N-1) \cdot (2N-1)}{6}$$

$$A = \sum_{m=1}^{N-1} A_m = \sum_{m=1}^{N-1} (N-m) \cdot (N-m) = \frac{N \cdot (N-1) \cdot (2N-1)}{6}$$

- Total Ops $= \dfrac{N \cdot (N-1) \cdot (4N+1)}{6} \propto \boxed{\dfrac{2}{3} N^3}$

Here we complete the operation counts for the Gaussian Elimination transformation. Since the mth column has (N-m) row elements below the diagonal that must be zeroed out, the total number of operations for the column is M_m = (N-m)*(N-m), A_m = (N-m)*(N-m), and D_m = 1 * (N-m) as shown. Summing over all columns except the last (which has no elements below the diagonal) we find the individual results for MAD and their sum yields the total number of operations as N(N-1)(4N+1)/6 which for large N is approximately (2/3) N^3. For N=1000, this yields nearly **one billion operations** (~(2/3) 10^9), so just think of the potential round off error! There are two types of sums to derive, namely, (i) $S_1 = \Sigma_{k=1}^N k$ and (ii) $S_2 = \Sigma_{k=1}^N k^2$. The first one $\Sigma_{k=1}^N k = 1+2+3+...+N$ can be evaluated as N/2 pairs each with the same sum (N+1); thus

$$S_1 = \Sigma_{k=1}^N k = (N/2)(N+1)$$

The second one $\Sigma_{k=1}^N k^2 = 1^2 + 2^2 + 3^2 + ... + N^2$ can be evaluated by the following reasoning: Since S_1, the sum over "k," is *quadratic in N*, we would expect S_2 the sum of "k^2," to be *cubic in N*, so try a solution of the form

$$S_2(N) = \Sigma_{k=1}^N k^2 = aN + bN^2 + cN^3$$

Next, writing this equation down for N= 1, 2, 3 (whose sums are easily computed) yields three equations in three unknowns, *viz.*,

$$S_2(N=1) = a(1) + b(1)^2 + c(1)^3 = 1^2 = 1$$
$$S_2(N=2) = a(2) + b(2)^2 + c(2)^3 = 1^2 + 2^2 = 5$$
$$S_2(N=3) = a(3) + b(3)^2 + c(3)^3 = 1^2 + 2^2 + 3^2 = 14$$

The solution is a=1/6, b=3/6, c=2/6, so we have $S_2(N) = (1/6)N + (3/6)N^2 + (2/6)N^3$ which can be written

$$S_2(N) = \Sigma_{k=1}^N k^2 = N(N+1)(2N+1)/6$$

6.3 Gaussian Elimination Example

Gaussian Elimination Example

$(E1):\ (-1)x_1 + 1x_2 - 4x_3 = 0$

$(E2):\ 2x_1 + (2)x_2 + 0x_3 = 1$

$(E3):\ 3x_1 + 3x_2 + (2)x_3 = 1/2$

$$\text{Total Ops} = \frac{N(N-1)(4N+1)}{6} = \frac{3(3-1)(4\cdot 3+1)}{6} = 13$$

Exclude ops on "b"

	M	A	D
Col1 Row2	2	2	1
Col1 Row3	2	2	1
Col2 Row3	1	1	1
Totals	5	5	3
MAD Total	13		

$$[A:b] = \begin{bmatrix} (-1) & 1 & -4 & \vdots & 0 \\ 2 & (2) & 0 & \vdots & 1 \\ 3 & 3 & (2) & \vdots & 1/2 \end{bmatrix}$$

$\rho_2 \Rightarrow \rho_2 - l_{21}\rho_1 = [2\ 2\ 0\ \vdots\ 1] - \left(\frac{2}{(-1)}\right)[-1\ 1\ -4\ \vdots\ 0] = [0\ 4\ -8\ \vdots\ 1]$

$\rho_3 \Rightarrow \rho_3 - l_{31}\rho_1 = [3\ 3\ 2\ \vdots\ 1/2] - \left(\frac{3}{(-1)}\right)[-1\ 1\ -4\ \vdots\ 0] = [0\ 6\ -10\ \vdots\ 1/2]$

$$[\tilde{A}:\tilde{b}] = \begin{bmatrix} (-1) & 1 & -4 & \vdots & 0 \\ 0 & (4) & -8 & \vdots & 1 \\ 0 & 6 & (-10) & \vdots & 1/2 \end{bmatrix}$$

$\rho_3 \Rightarrow \rho_3 - l_{32}\rho_2 = [0\ 6\ -10\ \vdots\ 1/2] - \left(\frac{6}{(4)}\right)[0\ 4\ -8\ \vdots\ 1] = [0\ 0\ 2\ \vdots\ -1]$

$$[U:c] = \begin{bmatrix} (-1) & 1 & -4 & \vdots & 0 \\ 0 & (4) & -8 & \vdots & 1 \\ 0 & 0 & (2) & \vdots & -1 \end{bmatrix} \begin{matrix} +5/4 \\ -3/4 \\ -1/2 \end{matrix}$$

$[0 - \{(1)\cdot(-3/4) + (-4)\cdot(-1/2)\}]/(-1)$ step #3

$[1 - (-8)\cdot(-1/2)]/(4)$ step #2

$-1/(2)$ step #1

U   c   x

From the system of three linear equations we easily identify the matrix **A** as well as the vector **b** whose components appear on the RHS of the three equations. Thus we write down the augmented matrix [**A**:**b**] and put parentheses around the diagonal elements to remind us that these elements are the "divisors".

1) We start the Gaussian Elimination procedure by zeroing out the rows of col#1 under the diagonal which are emphasized in grey in the matrix. The calculations are performed explicitly on the right of each row and are symbolized in the following suggestive manner:

$\rho_2 \rightarrow \rho_2 - l_{21}\rho_1$ with $l_{21} = a_{21}/a_{11} = 2/(-1)$, and

$\rho_3 \rightarrow \rho_3 - l_{31}\rho_1$ with $l_{31} = a_{31}/a_{11} = 3/(-1)$

This calculation yields the two replacement rows (including the new row values for the augmentation vector b) which appear in the new augmented matrix immediately below the original one.

2) Work now proceeds on col#2 which only has one element to zero out. That calculation appears to the right again symbolized as

$\rho_3 \rightarrow \rho_3 - l_{32}\rho_2$ with $l_{32} = a_{32}/a_{22} = 6/(4)$.

The replacement row appears in the final augmented matrix immediately below and yields the upper triangular matrix U

3) The augmented matrix [U:c] is now solved using backward elimination to yield the components of the solution vector **x** from the bottom up in the order x_3, then x_2, then x_1.

Note the Table in the upper right corner of the slide verifies the MAD operation formulas for the Gaussian Elimination algorithm by comparing the total directly to the actual count as we performed the calculation on the slide. In doing this we omitted the operations on the augmentation vector b as they were not included in our formulas. We will compute the additional operations involved in backward elimination later.

Direct Solution of Linear Systems

6.3.1 Revisit Gaussian Elimination

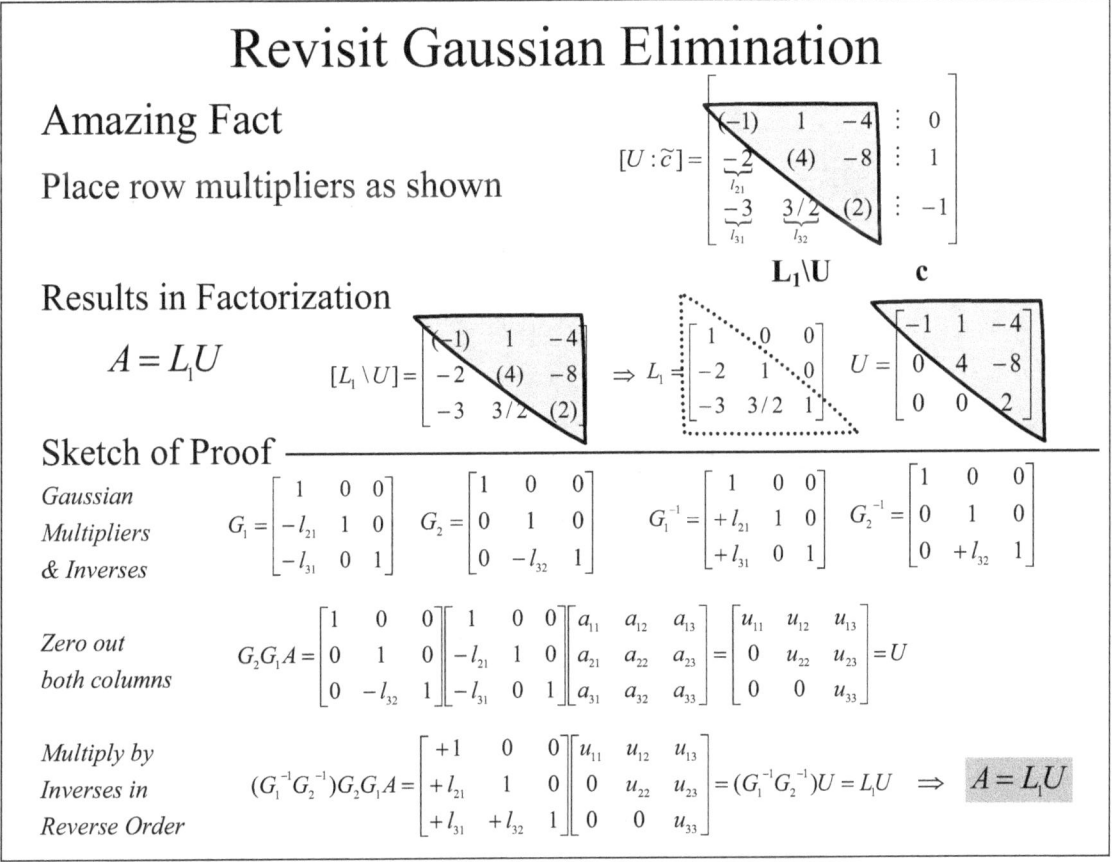

Revisiting the Gaussian elimination procedure, we can discover an interesting fact if we place the row multipliers in the appropriate row/col locations in the "empty" lower triangle. Specifically, we place $l_{21} = -2$ in row$_2$/col$_1$, $l_{31} = -3$ in row$_3$/col$_1$, and $l_{32} = 3/2$, in row$_3$/col$_2$, we obtain the matrix shown in the slide. The amazing thing is that if we identify these elements as those for a lower triangular matrix L_1 with *ones* placed along its diagonal, then the product of L_1 with the Gaussian upper triangular matrix U, produces the original matrix A, *i.e.*, we have the decomposition $A = L_1 U$. In other words Gaussian Elimination effectively decomposes the matrix A into the product of a "unit" lower triangular matrix L_1 and the upper triangular matrix U. It turns out that this is a very useful factorization of the matrix A as we discuss on the next slide.

The lower panel of the slide gives a sketch of the proof for a 3x3 matrix A. The Gaussian upper triangular matrices and their inverses are defined in terms of the Gaussian row multipliers l_{jk} on the 1st line. It is easily verified that $G_1^{-1} G_1 = G_2^{-1} G_2 = I$ as required for inverses. It is also easily verified that G_1 multiplying A "zeros out" the two entries under the diagonal in column #1, *viz.*,

$$-l_{21} \cdot a_{11} + a_{21} = -(a_{21}/a_{11}) \cdot a_{11} + a_{21} = 0 \quad ; \quad -l_{31} \cdot a_{11} + a_{31} = -(a_{31}/a_{11}) \cdot a_{11} + a_{31} = 0$$

Subsequent multiplication by G_2 then "zeros out" the one entry under the diagonal in column #2 resulting in an upper triangular matrix U as shown in the 2nd line. Finally a multiplication by the inverses in reverse order $G_1^{-1} \cdot G_2^{-1}$ leaves the matrix A on the LHS and $L_1 U$ on the RHS, which is the desired result.

6.3.2 Efficiency of $L_1\backslash U$ Factorization

Efficiency of $L_1\backslash U$ Factorization

- Define 2-step solution of $Ax = b$: $(L_1 \cdot \underbrace{U)x = b}_{\equiv c}$

  (1) $L_1 c = b$ solve for "c"

  (2) $Ux = c$ solve for "x"

- Advantage of 2-step procedure
  - Several systems of linear equations with same A
  $$Ax_1 = b_1 \;,\; Ax_2 = b_2 \;,\; \cdots,\; Ax_N = b_N$$
  - Only need "expensive" factorization $\propto \frac{2}{3}N^3$ **once**
  - Fwd & Bkwd substitutions $\propto 2N^2$

Gaussian Elimination leads to a factorization of the matrix $\mathbf{A} = \mathbf{L_1 U}$ which upon substitution into the matrix equation $\mathbf{Ax=b}$ yields a two step solution technique with some special advantages. Step #1 solves a lower triangular matrix equation $\mathbf{L_1 c = b}$ for an intermediate vector $\mathbf{c}$ and step#2 then solves an upper triangular matrix equation $\mathbf{U x = c}$ for the solution $\mathbf{x}$ to the original matrix equation.

At first glance it may seem that this is more work than just using Gaussian Elimination on the augmented matrix $[\mathbf{A:b}]$ to solve the original equation $\mathbf{A x = b}$ for $\mathbf{x}$. This is in fact true if there is only a single equation $\mathbf{A x = b}$ to be solved; however, it is often the case that we have a large number M of such equations to solve: $\mathbf{A x_1 = b_1}$, $\mathbf{A x_2 = b_2}$,..., $\mathbf{A x_M = b_M}$, where M can be as large as N and often larger for some problems.

If $\mathbf{A}$ is an N x N matrix, then each of the M Gaussian elimination solutions costs $\sim(2/3) N^3$ so the total cost is $\sim(2/3) N^3 M$ operations, which for the case M=N, is $\sim N^4$.

On the other hand, the two step procedure only requires $(2/3) N^3$ operations to create the factorization of $\mathbf{A = L_1 U}$ once ; after that, the solution of M the forward and M backward equations requires $\sim 2N^2$ M operations so that the total MAD operations is $(2/3) N^3 + 2N^2 M$, which for the case M=N, is $\sim N^3$.

Clearly, the bulk of the work is in factoring the matrix $\mathbf{A}$ and therefore the advantage of the two-step method is that it performs this only once, while Gaussian elimination must do it for each of the M equations.

6.3.3 Counting: Forward/Backward Elimination

Counting: Forward/Backward Elimination

- **Bkwd** $x_k = \{c_k - \sum_{\alpha=k+1}^{N} u_{k\alpha} x_\alpha\} / u_{kk}$

  $\underbrace{M=(N-k)}_{A=1}, \underbrace{A=(N-k-1)}_{}, D=1$

  $M = A = \sum_{k=1}^{N}(N-k) = \dfrac{N\cdot(N-1)}{2}$

  $D = N$

  $M + A + D = \dfrac{N\cdot(N-1)}{2} + \dfrac{N\cdot(N-1)}{2} + N = N^2$

- **Fwd** $c_k = \{b_k - \sum_{\alpha=1}^{k-1} l_{k\alpha} c_\alpha\} / l_{kk}$

  $\underbrace{M=(k-1)}_{A=1}, \underbrace{A=(k-2)}_{}, D=0,\ l_{kk}=1$

  $M = A = \sum_{k=1}^{N}(k-1) = \dfrac{N\cdot(N-1)}{2}$

  $D = 0$ (divide by 1 unnecessary)

  $M + A + D = N\cdot(N-1)$

- **Fwd + Bkwd** $M + A + D = 2N^2 - N \propto 2N^2$

- **$L_1 \backslash U$ Factorization** $M + A + D = \dfrac{N\cdot(N-1)\cdot(4N+1)}{6} \propto \dfrac{2}{3}N^3$

- **Both** $M + A + D \propto \underbrace{\dfrac{2}{3}N^3}_{L_1\backslash U} + \underbrace{2N^2}_{Bkwd\,\&\,Fwd}$

  Sum Formulas: $\sum_{k=1}^{n} k = n(n+1)/2$, $\sum_{k=1}^{n} k^2 = n(n+1)(2n+1)/6$

The MAD operation count for the Backward and Forward Solutions are calculated from the general formulas previously given.

BKWD: The index $k =1,2,...,N$ and the equation for a fixed index has a sum of products running over the range $\alpha = k+1$ to N corresponding to $M = N-k$ products and one less add $A= N-k-1$; there is one additional add within the braces ("c_k – sum") bringing the total number of adds to $A= N-k$; finally there is one divide of the braced term by u_{kk}, so $D=1$. Summing $N-k$ from $k = 1$ to N yields both multiply and add operations $M=A=N(N-1)/2$ and summing $D = 1$ from $k=1$ to N gives $D=N$; finally summing all three yields N^2 operations for the backward solution.

FWD: The index $k =1,2,...,N$ and the equation for a fixed index has a sum of products running over the range $\alpha = 1$ to $k-1$ corresponding to $M = (k-1)$ products and one less add $A= (k-2)$; again there is one additional add within the braces ("b_k – sum") bringing the total number of adds to $A= (k-1)$; finally there is one divide of the braced term by $l_{kk} =1$ which is unnecessary and makes $D=0$. Performing the identical sums for M and A yields a total number of operations $2*N(N-1)$ which is $\sim 2N^2$.

The total number of operations for the two-step solution to the matrix equation $\mathbf{Ax} = \mathbf{b}$ is the sum of the factorization $L_1 \backslash U$ and backward and forward solution MAD operations as shown in the last boxed equation MAD $\sim (2/3) N^3 + 2 N^2$.

Direct Solution of Linear Systems

6.3.4 Efficient Methods for Special Types of Matrices

Efficient Methods for Special Types of Matrices

- **Banded Matrices**
  - Tri-diagonal

$$T = \begin{bmatrix} l_1 & d_1 & u_1 & 0 & 0 & 0 \\ & l_2 & d_2 & u_2 & 0 & 0 \\ & 0 & \ddots & \ddots & \ddots & 0 \\ & 0 & 0 & l_{N-1} & d_{N-1} & u_{N-1} \\ & 0 & 0 & 0 & l_N & d_N & u_N \end{bmatrix} = [\vec{l}^T, \vec{d}^T, \vec{u}^T]$$

$$\vec{l} = [\underbrace{l_1}_{=0}, l_2, \cdots, l_{N-1}, l_N]$$

$$\vec{d} = [d_1, d_2, \cdots, d_{N-1}, d_N]$$

$$\vec{u} = [u_1, u_2, \cdots, u_{N-1}, \underbrace{u_N}_{=0}]$$

  - Special Gaussian Algorithm $\quad M + A + D = (8N - 7) \propto N$
  - Huge advantage over full Gaussian $\quad \propto \frac{2}{3} N^3$
  - Toeplitz, Circulant, Vandermonde, Hilbert, Block Diagonal, Sparse.

We have seen that the total number of operations for the two-step solution to the matrix equation $\mathbf{Ax} = \mathbf{b}$ is the sum MAD ~ $(2/3) N^3 + 2 N^2$. This method is costly and must be used if all elements of the N x N matrix $\mathbf{A}$ are non-zero; however there are other methods specifically tailored to special types of matrices having special structure, symmetries, or have very few non-zero elements. These matrices go by names such as Tri-diagonal, Banded, Toeplitz, Vandermonde, Hilbert, Block Diagonal, Sparse, to mention a few.

As an example, the Banded matrices are defined by their bandwidth or the number of sub-diagonals there are above and below the main diagonal. The Tri-diagonal case consisting of the N main diagonal elements $\{d_1, d_2, ..., d_N\}$ together with (N-1) upper $\{u_1, u_2, ..., u_{N-1}\}$ and (N-1) lower $\{l_2, l_3, ..., l_N\}$ elements along the sub-diagonals is illustrated in the figure. The Tri-diagonal matrix has a total of 3N-2 elements rather than the full complement of N^2 and a special algorithm can be constructed for a Tri-diagonal martix equation $\mathbf{T_3 x} = \mathbf{b}$ which requires only a total of (8N-7) MAD operations for its solution. For large N matrices, this method, with a computation cost ~8N, has a huge advantage over the full Gaussian solution which has a computational cost ~$(2/3) N^3$. Thus, whenever possible, these special methods should be sought out and used.

6.4 The Need for Matrix Pivoting

Why Pivoting?

- **Need for pivoting**
  - Gaussian Elimination fails for zero diagonals $a_{mm} = 0$
  - Round-off error $a_{mm} \sim 0$
    - Large row multiplier $l_{km} = a_{km} / a_{mm}$
    - Small divisor $a_{mm} \sim 0$ in back substitution

- **Example:**
$$.003 x_1 + 59.14 x_2 = 59.17$$
$$5.291 x_1 - 6.130 x_2 = 46.78$$

$$x_1^{exact} = +10.00$$
$$x_2^{exact} = +1.000$$

$$\begin{bmatrix} .003 & 59.14 & \vdots & 59.17 \\ 5.291 & -6.130 & \vdots & 46.78 \end{bmatrix}$$
$$\quad A \qquad\qquad b$$

Zero 1st Col. $\rho_2 \Rightarrow [5.291 \;\; -6.130 \;\; 46.78] - \dfrac{5.291}{.003} \cdot [.003 \;\; 59.14 \;\; 59.17] = [0 \;\; -104{,}300 \;\; -104{,}400]$

$$\begin{bmatrix} .003 & 59.14 & \vdots & 59.17 & \vdots & -10.00 \\ 0 & -104{,}300 & \vdots & -104{,}400 & \vdots & 1.001 \end{bmatrix}$$
$$\quad \tilde{A} \qquad\qquad \tilde{b} \qquad\qquad x$$

$x_1 = [59.17 - \overbrace{59.14 \cdot 1.001}^{=59.20}] / .003$

$x_2 = (-104{,}400)/(-104{,}300)$

x_1 has wrong sign!

In the introductory remarks we gave a simple example in which the direct solution to the matrix equation **A x = b** led to a solution that had lost significant digits because of numerical round off error. The example on this slide shows a case in which the straight forward Gaussian Elimination procedure gives a solution that not only loses all 4 of its significant digits, but also has the wrong sign. Following the explicit calculation we see that dividing by the small diagonal term $a_{11} = .003$ is the culprit that leads to this disastrous result.

For the general case in N-dimensions, this problem may occur every time we divide by a small diagonal element a_{mm} and the effects of these round off errors will propagate throughout the Gaussian factorization and the Backward/Forward solution procedures, thereby losing significant digits in each component of the solution. Thus we need some procedure to prevent us from dividing by small diagonal terms; for the two dimensional case shown, this may be accomplished by simply swapping the two rows so that the second row now contains the small term .003 which is no longer in a diagonal position and hence no longer a "divisor" after the row swap.

The procedure of swapping rows in order to move small elements off the diagonal is called "pivoting" and is equivalent to solving the equations in a different order. In the next few slides we shall see how pivoting effectively takes care of this numerical rounding problem.

6.4.1 Why Scaled Pivoting?

Why Scaled Pivoting

- Solve Example with partial pivoting
- Pivot: 1st col. Max element in row 2: $\rho_1 \longleftrightarrow \rho_2$

$$\begin{bmatrix} .003 & 59.14 & \vdots & 59.17 \\ 5.291 & -6.130 & \vdots & 46.78 \end{bmatrix} \xrightarrow{1 \leftrightarrow 2} \begin{bmatrix} 5.291 & -6.130 & \vdots & 46.78 \\ .003 & 59.14 & \vdots & 59.17 \end{bmatrix}$$

- Zero 1st Col. $\rho_2 \Rightarrow [.003 \ 59.14 \ 59.17] - \dfrac{.003}{5.291} \cdot [5.291 \ -6.130 \ 46.78] = [0 \ 59.14 \ 59.14]$

- Solve
$$\begin{bmatrix} 5.291 & -6.130 & \vdots & 46.78 & 10.00 \\ 0 & 59.14 & \vdots & 59.14 & 1.000 \end{bmatrix} \quad \begin{array}{l} \overbrace{x_1 = [46.78 - (-6.130) \cdot 1.000]}^{=52.91}/5.291 \\ x_2 = 59.14/59.14 \end{array}$$

$\tilde{A} \qquad \tilde{b} \qquad x$

- Same example with (1st row) * 10^4 $\begin{bmatrix} 30.00 & 591{,}400 & \vdots & 591{,}700 \\ 5.291 & -6.130 & \vdots & 46.78 \end{bmatrix}$
- PP does not yield pivot!
  - Define row scale vector $\quad \vec{s} = \begin{bmatrix} 591{,}400 \\ 6.130 \end{bmatrix}$
  - Col. 1 scaled elements: $\begin{bmatrix} 30.00/591{,}400 \\ 5.291/6.130 \end{bmatrix} \longleftarrow$ Large
  - Thus, pivot $\rho_1 \longleftrightarrow \rho_2$

The numerical rounding issues of the previous example are "cured" by swapping rows 1 and 2 because the new augmented matrix has the largest element "5.291" at the diagonal position. Proceeding with Gaussian Elimination we zero out the 1st column below the diagonal by the row transformation shown obtain the **U** matrix shown. The computations carried out to the right of the augmented matrix first yield $x_2 = 1.000$ and then yield $x_1 = 10.00$ both of which are correct to 4 significant digits.

In the lower panel we consider the same set of equations, but we multiply the first equation by 10^4; this changes the first row of the augmented matrix to [30.00 591,400 : 591700] as shown. Now searching down the first column shows the maximum element to be 30.00 and Partial Pivoting criterion would tell us not to swap since the maximum element is already on the diagonal. The floating point representation maintains significant digits when we scale the exponent up or down, so clearly, we have not changed anything numerically; yet PP tells us not to swap rows as we know we should. Solving this problem as it stands leads to the same incorrect answers as on the previous slide. The solution is to write down a scale vector consisting of the largest element of **A** in each row and then compare the **scaled values** in column #1 to determine whether or not to swap rows. The row scale vector s = [591400 , 6.130]^T and the scaled elements of col#1 become
$$[\ 30.00/591400\ ,\ 5.291/6.130]^T = [\ .00005\ ,\ .8631]^T$$
so we should swap rows 1 and 2.

6.4.2 Pivoting Overview

> # Pivoting Overview
>
> - Avoid small divisors by maximizing diagonal elements
> - Perform row interchanges *prior to zeroing column*
> - Three Types of Pivoting
>   - They all find a maximum element a_{pm} in *col.m*
>   - Interchange entire rows of augmented matrix [**A**:**b**]
> $$\rho_p \leftrightarrow \rho_m$$
>   - Zero under *col. m* and proceed to next *col. m+1*
>   - Differ in method of selection of pivot row
>   1. **Partial Pivoting**: Directly find max element a_{pm} in *col.m*
>   2. **Scaled Partial Pivoting**: First form a row scale vector **s** containing max absolute value of elements in each row; then compare *scaled elements* in *col. m* to find max scaled row (a_{pm}/s_p); carry scale vector in augmented matrix [**A**:**b**:**s**]
>   3. **Scaled Full Pivoting**: similar to 2. above but apply scaling to whole submatrix A_{mm} ; exchange both rows and columns

In order to avoid small divisors a "pivoting" or row-swapping procedure is used during Gaussian Elimination in order to place large elements in the diagonal position prior to zeroing out a column. There are three types of pivoting as summarized in the slide and detailed below

(i) Partial Pivoting (PP) looks down the "target" column "m" to find its largest element (absolute value) a_{pm} and then swaps row_p with row_m (entire row of augmented matrix [**A**: **b**]). Once done, the largest element becomes the divisor for that column and the Gaussian Elimination steps are now performed to zero out all the elements below the diagonal. For the next column "m+1" we repeat this procedure again swapping rows to bring the largest element in the column up to the diagonal position for that column and then proceeding to zero out the column below the new diagonal element $a_{m+1,m+1}$.

(ii) Scaled Partial Pivoting (SPP) first establishes a scale vector $\mathbf{s} = [s_1, s_2, ...s_m, ..., s_N]^T$ whose components are the largest (absolute value) elements in each row of the matrix **A** . Next we create a temporary scaled column vector $[a_{mm}/s_m, a_{m-1,m}/s_{m-1}, ..., a_{N,m}/s_N]^T$ in which every element a_{im} in column "m" (below the diagonal) is divided by its scale vector component s_i . Now the row swap is determined by finding the maximum "scaled component" a_{pm}/s_p . Then, just as in PP we swap row_p with row_m and proceed to zero out the elements below the diagonal. In order to keep row scales properly aligned, we must also swap the components of the scale vector; this is done by "carrying" the scale vector along with the augmented matrix in the structure [**A**:**b**:**s**] so all swaps are made simultaneously. We continue with this procedure as in PP always using the original scale vector (with row swapped as we proceed).

(ii) Scaled Full Pivoting (SFP) is similar to SPP in its use of a single scale vector s, but instead of just searching down the column for the maximum element a_{pq} a "full" search through the entire "scaled" sub-matrix A_{mm} is done and then the maximum element a_{pq} is migrated to the head of the column by exchanging both rows and columns. We shall motivate the logic behind these three methods as we take a more detailed look in the following few slides.

6.4.3 Partial Pivoting

Partial Pivoting

1. Partial Pivoting:

Make diagonal elements larger than any elements below it prior to zeroing

a) Thus if a_{pm} is max element in *col. m*, interchange *row* ρ_p with ρ_m so that new diagonal element is largest

b) Then zero *col. m* below diagonal and continue with *col. m+1*

c) This insures that we **divide by** the **largest number in each column** when solving the linear system **Ax=b** thereby avoiding RO problems that are amplified when dividing by a small number.

Partial Pivoting is the least computationally costly method of avoiding small divisors and is therefore the best choice if the matrix **A** has elements that are all of the same order of magnitude. At the stage illustrated in the diagram a search is performed down the "target" column "m" to find its largest element (absolute value) a_{pm} and then the row swap is made between row$_p$ and row$_m$ (entire row of augmented matrix [**A**: **b**]). Once done, the largest element becomes the divisor for that column and the Gaussian Elimination steps are now performed to zero out all the elements below the diagonal. For the next column "m+1" we repeat this procedure again swapping rows to bring the largest element in the column up to the diagonal position for that column and then proceeding to zero out the column below the new diagonal element $a_{m+1,m+1}$.

6.4.4 Scaled Pivoting

Scaled Pivoting

2. Scaled Partial Pivot

a) Form a scale vector containing largest absolute value in each row (once)

b) Find max scaled value a_{pm} / s_p and interchange *row p* with *m*

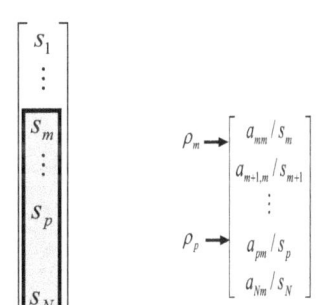

3. Scaled Full Pivot

a) Search all elements in sub-matrix $\mathbf{A}_{mm}$ for the largest scaled value and

b) Interchange rows and columns to migrate a_{pq} the maximum scaled element to the a_{mm} position

$$\begin{matrix} & & & & \text{Col. } q \\ & & & & \downarrow \\ \rho_m & \begin{bmatrix} \boxed{a_{mm}/s_m} & a_{m,m+1}/s_m & \cdots & a_{mN}/s_m \\ \vdots & \vdots & & \vdots \\ \rho_p & a_{pm}/s_p & a_{p,m+1}/s_p & \boxed{a_{p,q}/s_p} & a_{iN}/s_p \\ a_{Nm}/s_N & a_{N,m+1}/s_N & \cdots & a_{NN}/s_N \end{bmatrix} \end{matrix}$$

sub-matrix $\mathbf{A}_{mm}$

Scaled Partial Pivoting requires an initial search across the whole matrix in order to establish the scale vector "**s**" and thus is more computationally expensive than Partial Pivoting. Scaling is required whenever the matrix **A** has elements that are not the same order of magnitude because a simple search for the maximum in the target column would find the row with all large elements as the pivot row. Since any row could be made arbitrarily large or small and still not change the system of equations, such pivoting without scaling cannot find the truly dominant element. Scaled Full Pivoting is similar to the partial pivoting but entails a search of the entire sub matrix for each column and therefore has the largest computational cost of the three methods discussed; it would appear to be a choice for only the most difficult numerical situations.

Scaled Partial Pivoting (SPP) At the stage illustrated in the upper diagram a search is performed down the "target" column "m" to find the largest element of the scaled column vector $[\,a_{mm}/s_m\,,\,a_{m-1,m}/s_{m-1}\,,...,\,a_{N,m}/s_N\,]^T$; we then perform the swap of row$_p$ with row$_m$ and proceed to zero out the elements below the diagonal using the new diagonal element. The scale vector $\mathbf{s} = [s_1, s_2, ...s_m, ..., s_N]^T$ whose components are the largest (absolute value) elements in each row of the matrix **A** is determined once at the beginning and then carried along as part of the augmented matrix structure [**A:b:s**] so all swaps are made simultaneously.

Scaled Full Pivoting (SFP) At the stage illustrated in the lower diagram a search is performed across the entire sub-matrix $\mathbf{A}_{mm}$ (corresponding to the "target" column "m") to find the largest element of the scaled sub-matrix $\mathbf{A}_{mm}$ and then the maximum element $a_{p,q}$ is migrated to the head of the column by exchanging both rows and columns. We then proceed to zero out the elements below the diagonal in column "m" using the new diagonal element on the resulting **A** matrix. The next column "m+1" is handled in a similar manner by considering the scaled sub-matrix $\mathbf{A}_{m+1,m+1}$ and once again migrating the maximum element via row and column swaps and then proceeding to zero out column "m+1" below the diagonal. Clearly, the computational cost of SFP is the largest of the three methods and is only used when the other do not work well.

6.5 Direct Matrix Inversion

Direct Matrix Inversion

- Solve multiple problems at once and choose **b** s appropriately

$$\mathbf{Ax}_1 = \mathbf{b}_1 \qquad \begin{bmatrix} a_{11} & a_{12} & a_{13} \\ a_{21} & a_{22} & a_{23} \\ a_{31} & a_{32} & a_{33} \end{bmatrix} \begin{bmatrix} \bar{x}_1)_1 \\ \bar{x}_1)_2 \\ \bar{x}_1)_3 \end{bmatrix} = \begin{bmatrix} 1 \\ 0 \\ 0 \end{bmatrix}$$

$$\mathbf{Ax}_2 = \mathbf{b}_2 \qquad \begin{bmatrix} a_{11} & a_{12} & a_{13} \\ a_{21} & a_{22} & a_{23} \\ a_{31} & a_{32} & a_{33} \end{bmatrix} \begin{bmatrix} \bar{x}_2)_1 \\ \bar{x}_2)_2 \\ \bar{x}_2)_3 \end{bmatrix} = \begin{bmatrix} 0 \\ 1 \\ 0 \end{bmatrix}$$

$$\mathbf{Ax}_3 = \mathbf{b}_3 \qquad \begin{bmatrix} a_{11} & a_{12} & a_{13} \\ a_{21} & a_{22} & a_{23} \\ a_{31} & a_{32} & a_{33} \end{bmatrix} \begin{bmatrix} \bar{x}_3)_1 \\ \bar{x}_3)_2 \\ \bar{x}_3)_3 \end{bmatrix} = \begin{bmatrix} 0 \\ 0 \\ 1 \end{bmatrix}$$

- Adjoin columns $A[\bar{x}_1 \vdots \bar{x}_2 \vdots \bar{x}_3] = [\bar{b}_1 \vdots \bar{b}_2 \vdots \bar{b}_3]$
  Solve matrix eqn 1 col. at a time $\mathbf{AX}_k = \mathbf{B}_k$

$$\underbrace{\begin{bmatrix} a_{11} & a_{12} & a_{13} \\ a_{21} & a_{22} & a_{23} \\ a_{31} & a_{32} & a_{33} \end{bmatrix}}_{\mathbf{A}} \underbrace{\begin{bmatrix} \bar{x}_1)_1 \\ \bar{x}_1)_2 \\ \bar{x}_1)_3 \end{bmatrix} \begin{bmatrix} \bar{x}_2)_1 \\ \bar{x}_2)_2 \\ \bar{x}_2)_3 \end{bmatrix} \begin{bmatrix} \bar{x}_3)_1 \\ \bar{x}_3)_2 \\ \bar{x}_3)_3 \end{bmatrix}}_{\mathbf{X}} = \underbrace{\begin{bmatrix} 1 & 0 & 0 \\ 0 & 1 & 0 \\ 0 & 0 & 1 \end{bmatrix}}_{\mathbf{B} = \mathbf{I}_3}$$

- Col-by-col. soln for inverse $\mathbf{X} = \mathbf{A}^{-1}$
  (Gaussian Elimination)

$$\mathbf{A} \cdot \mathbf{A}^{-1} = \mathbf{I}$$
$$\mathbf{A} \cdot [\mathbf{A}^{-1}]_k = \mathbf{I}_k$$
$$\mathbf{A} \cdot col_k[\mathbf{A}^{-1}] = col_k[\mathbf{I}]$$

- Generalizes to N-dimensions

The numerical inverse of an N x N matrix **A** can be computed directly by solving a set of N matrix equations of the form $\{\mathbf{A}\ \mathbf{x}_1 = \mathbf{b}_1, \mathbf{A}\ \mathbf{x}_2 = \mathbf{b}_2, ..., \mathbf{A}\ \mathbf{x}_N = \mathbf{b}_N\}$ provided we choose the "b"s appropriately. The choices for N=3 are shown on the slide; the three x vectors have scalar components given by the column vectors

$$\mathbf{x}_1 = [x_{1,1}, x_{1,2}, x_{1,3}]^T, \qquad \mathbf{x}_2 = [x_{2,1}, x_{2,2}, x_{2,3}]^T, \qquad \mathbf{x}_3 = [x_{3,1}, x_{3,2}, x_{3,3}]^T$$

and when juxtaposed these three column vectors form a partitioned matrix $X = [\mathbf{x}_1 : \mathbf{x}_2 : \mathbf{x}_3]$
Similarly the three b vectors have scalar components given by the column vectors

$$\mathbf{b}_1 = [1, 0, 0]^T, \qquad \mathbf{b}_2 = [0, 1, 0]^T, \qquad \mathbf{b}_1 = [0, 0, 1]^T,$$

and when juxtaposed these three column vectors form a partitioned matrix

$$\mathbf{B} = [\mathbf{b}_1 : \mathbf{b}_2 : \mathbf{b}_3] = diag(1,1,1) = \mathbf{I}$$

In this column-vector form, the three matrix equations $\mathbf{A}\ \mathbf{x} = \mathbf{b}$ appear as the single matrix product equation

$$\mathbf{A}\ \mathbf{X} = \mathbf{B} = \mathbf{I}$$

But this last equation defines **X** to be the inverse of **A**. Thus the inverse **X** can be found by solving the partitioned matrix one column vector at a time and the juxtaposition of these column vectors forms a partitioned matrix that is precisely the inverse of **A**.

Direct Solution of Linear Systems

6.5.1 Gauss Matrix Inversion

Gauss Matrix Inversion

- **Gauss L\U:** $[\mathbf{A}:\mathbf{I}] \xrightarrow{\text{Triangular ize}} [\mathbf{L}\backslash\mathbf{U}:\tilde{\mathbf{I}}] \xrightarrow{\text{Backsolve}} [\mathbf{L}\backslash\mathbf{U}:\tilde{\mathbf{I}}:\mathbf{A}^{-1}]$

- **Gauss Example**

$$\begin{bmatrix} (1) & 2 & -1 & : & 1 & 0 & 0 \\ 2 & (1) & 0 & : & 0 & 1 & 0 \\ -1 & 1 & (2) & : & 0 & 0 & 1 \end{bmatrix}$$
$$\mathbf{A} \qquad \mathbf{I}$$

Col.# 1: Rows 2 & 3

$\rho_2 \to \rho_2 - \dfrac{a_{21}}{a_{11}}\cdot\rho_1 = [2\ 1\ 0\ :\ 0\ 1\ 0] - \dfrac{2}{(1)}\cdot[1\ 2\ -1\ :\ 1\ 0\ 0]$ $l_{21}\rho_1$

$= [0\ (-3)\ 2\ :\ -2\ 1\ 0]$

$\rho_3 \to \rho_3 - \dfrac{a_{31}}{a_{11}}\cdot\rho_1 = [-1\ 1\ 2\ :\ 0\ 0\ 1] - \dfrac{-1}{(1)}\cdot[1\ 2\ -1\ :\ 1\ 0\ 0]$ $l_{31}\rho_1$

$= [0\ (3)\ 1\ :\ 1\ 0\ 1]$

$$\begin{bmatrix} (1) & 2 & -1 & : & 1 & 0 & 0 \\ 0 & (-3) & 2 & : & -2 & 1 & 0 \\ 0 & 3 & (1) & : & 1 & 0 & 1 \end{bmatrix}$$
$$\tilde{\mathbf{A}} \qquad \tilde{\mathbf{I}}$$

Col.# 2: Row 3

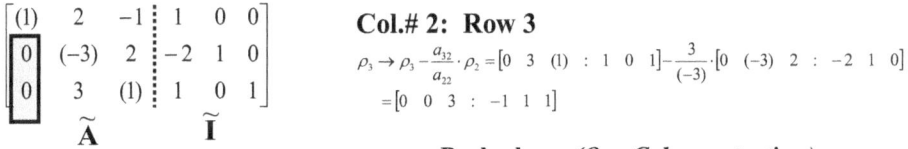

$= [0\ 0\ 3\ :\ -1\ 1\ 1]$ $l_{32}\rho_2$

Backsolve   *(One Column at a time)*

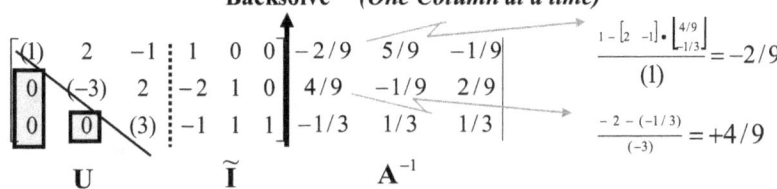

An explicit calculation of the inverse of the matrix **A** using Gaussian Elimination on the partitioned **B** matrix is detailed on this slide. The matrix of juxtaposed **b** vectors **B** is just the identity matrix I = diag(1,1,1) and the 3 x 6 augmented matrix [**A**: **I**] can be thought of as solving three separate **Ax=b** problems as previously discussed. In this way, the row replacements of the Gaussian elimination process are just extended to include all three columns of **I** simultaneously so for example when zeroing out the sub-diagonal in the first column we have

$\text{row}_2 = \text{row}_2 - l_{21}\,\text{row}_1 = [2\ 1\ 0\ :\ 1\ 0\ 0] - \{2/(1)\}[1\ 2\ -1\ :\ 1\ 0\ 0] = [0\ -3\ 2\ :\ -2\ 1\ 0]$

$\text{row}_3 = \text{row}_3 - l_{31}\,\text{row}_1 = [-1\ 1\ 2\ :\ 0\ 0\ 1] - \{-1/(1)\}[1\ 2\ -1\ :\ 1\ 0\ 0] = [0\ 3\ 1\ :\ 1\ 0\ 1]$

Continuing in the same manner for the 2nd column we obtain the 3 x 6 augmented matrix [**U**: **I-tilde**] shown on the bottom of the slide. We backsolve this equation using one column of I-tilde at a time ; thus backsolving [**U**:col$_1$(**I-tilde**)] yields the components of the 1st column: col$_1$(**A**$^{-1}$) from bottom to top as indicated by the dark arrow pointing up. In the same manner, backsolving [**U**:col$_2$(**I-tilde**)] yields the components of the 2nd column: col$_2$(**A**$^{-1}$) ; and backsolving [**U**:col$_3$(**I-tilde**)] yields the components of the 3rd column: col$_3$(**A**$^{-1}$) completing the solution for all elements of the inverse **A**$^{-1}$. method which continues the Gaussian Elimination

Direct Solution of Linear Systems

6.5.2 Gauss-Jordan Matrix Inversion

Gauss-Jordan Matrix Inversion

Gauss-Jordan $[A:I] \xrightarrow{\text{Diagonalize}} [D:\tilde{I}] \xrightarrow{\text{Divide rows by } d_k} [I:A^{-1}]$

- **Gauss-Jordan Example**   Col.# 1: Rows 2 & 3 Identical to Gauss

Col.# 2: Row 1 & 3

$$\begin{bmatrix} (1) & 2 & -1 & : & 1 & 0 & 0 \\ 0 & (-3) & 2 & : & -2 & 1 & 0 \\ 0 & 3 & (1) & : & 1 & 0 & 1 \end{bmatrix}$$
$\mathbf{A}_1 \qquad \tilde{\mathbf{I}}$

$\rho_1 \to \rho_1 - \frac{a_{12}}{a_{22}} \cdot \rho_2 = [(1)\ 2\ -1\ :\ 1\ 0\ 0] - \frac{2}{(-3)} \cdot [0\ (-3)\ 2\ :\ -2\ 1\ 0]$
$= [1\ 0\ 1/3\ :\ -1/3\ 2/3\ 0]$

$\rho_3 \to \rho_3 - \frac{a_{32}}{a_{22}} \cdot \rho_2 = [0\ 3\ (1)\ :\ 1\ 0\ 1] - \frac{3}{(-3)} \cdot [0\ (-3)\ 2\ :\ -2\ 1\ 0]$
$= [0\ 0\ 3\ :\ -1\ 1\ 1]$

Col.# 3: Row 1 & 2

$$\begin{bmatrix} (1) & 0 & 1/3 & : & -1/3 & 2/3 & 0 \\ 0 & (-3) & 2 & : & -2 & 1 & 0 \\ 0 & 0 & (3) & : & -1 & 1 & 1 \end{bmatrix}$$
$\mathbf{A}_2 \qquad \tilde{\mathbf{I}}_2$

$\rho_1 \to \rho_1 - \frac{a_{13}}{a_{33}} \cdot \rho_3 = [(1)\ 0\ 1/3\ :\ -1/3\ 2/3\ 0] - \frac{1/3}{(3)} \cdot [0\ 0\ 3\ :\ -1\ 1\ 1]$
$= [1\ 0\ 0\ :\ -2/9\ 5/9\ -1/9]$

$\rho_2 \to \rho_2 - \frac{a_{23}}{a_{33}} \cdot \rho_3 = [0\ (-3)\ 2\ :\ -2\ 1\ 0] - \frac{2}{(3)} \cdot [0\ 0\ 3\ :\ -1\ 1\ 1]$
$= [0\ -3\ 0\ :\ -4/3\ 1/3\ -2/3]$

$$\begin{bmatrix} (1) & 0 & 0 & : & -2/9 & 5/9 & -1/9 \\ 0 & (-3) & 0 & : & -4/3 & 1/3 & -2/3 \\ 0 & 0 & (3) & : & -1 & 1 & 1 \end{bmatrix}$$
$\mathbf{D} \qquad \tilde{\mathbf{I}}_3$

Divide
row 1 by 1
row 2 by –3
row 3 by 3

$\longrightarrow$

$$\begin{bmatrix} (1) & 0 & 0 & : & -2/9 & 5/9 & -1/9 \\ 0 & (1) & 0 & : & +4/9 & -1/9 & +2/9 \\ 0 & 0 & (1) & : & -1/3 & 1/3 & 1/3 \end{bmatrix}$$
$\mathbf{I} \qquad \mathbf{A}^{-1}$

An alternate calculation of the inverse of the matrix **A** uses the Gauss-Jordan method which continues the Gaussian Elimination by zeroing out the terms above the diagonal as well. The steps taken to zero out terms under the 1st column are identical to Gaussian Method; the 2nd column row$_3$ calculation is also unchanged, but row$_1$ above the diagonal has to be zeroed out as well

row$_1$ = row$_1$ − l_{12} row$_2$ = [1 2 -1 : 1 0 0] − {2/(-3)}[0 -3 2 : -2 1 0] = [1 0 1/3 : -1/3 2/3 0]
row$_3$ = row$_3$ − l_{32} row$_2$ = [0 3 1 : 1 0 1] − {3/(-3)}[0 -3 2 : -2 1 0] = [0 0 3: -1 1 1]

The 3rd column requires zeros in row$_1$ and row$_2$

row$_1$ = row$_1$ − l_{13} row$_3$ = [1 0 1/3 : -1/3 2/3 0] − {(1/3)/(3)} [0 0 3: -1 1 1] = [1 0 0 : -2/9 5/9 -1/9]
row$_2$ = row$_2$ − l_{23} row$_3$ = [0 -3 2: -2 1 0] − {2/(3)} [0 0 3: -1 1 1] = [0 -3 0 : -4/3 1/3 -2/3]

This leaves the 3 x 6 augmented matrix with $\mathbf{A}_3 = \mathbf{D}$ = diag(1, -3,3) and I$_3$-tilde: [**D**: I$_3$-tilde]
The resulting inverse just requires division of each row of the I$_3$-tilde matrix by the diagonal elements {1,-3, 3} of **D** to yield $\mathbf{A}^{-1}$ as shown

$$A^{-1} = \begin{bmatrix} -2/9 & 5/9 & -1/9 \\ 4/9 & -1/9 & 2/9 \\ -1/3 & 1/3 & 1/3 \end{bmatrix}$$

6.6 3^d Determinants

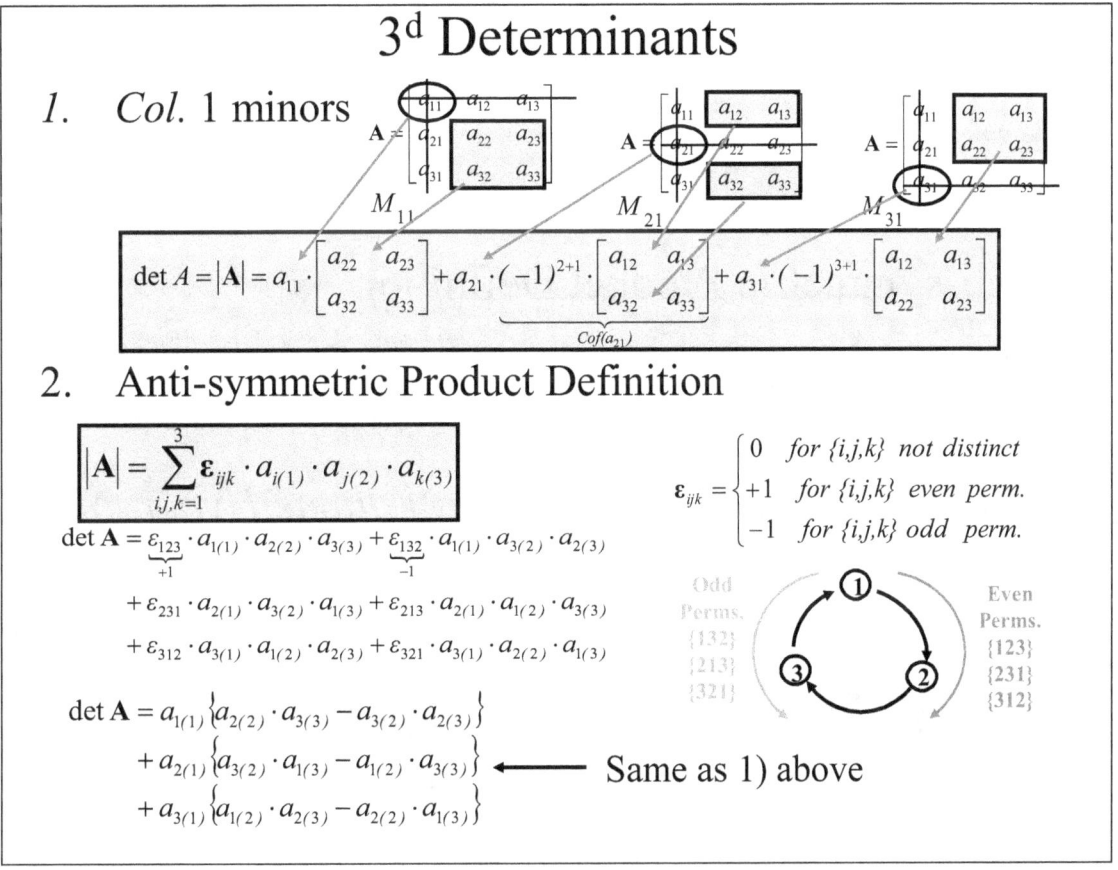

The determinant of a 3 x 3 matrix is displayed in detail as an expansion by "minors" of col_1. The a_{11} minor is found by striking row_1 and col_1 leaving the 2x2 minor determinant shown multiplied by a_{11}; similarly, the a_{21} minor is found by striking row_2 and col_1 leaving the 2nd 2x2 minor determinant shown multiplied by a_{21}; finally, the a_{31} minor is found by striking row_3 and col_1 leaving the 3rd 2x2 minor determinant shown multiplied by a_{31}. Note that the sign alternates from term to term as $(-1)^{r+c}$ and this is often attached to the minor and called the cofactor

$$\text{cof}(a_{rc}) = (-1)^{r+c} \text{minor}(a_{rc})$$

This expansion is easily extended to higher dimensions and can also be expanded using any row or column that provides an efficient computation (*i.e.*, the row or column with the most zeros in it).

An alternate method of computing the completely anti-symmetric product of triples of matrix elements multiplied by the "alternating symbol" ε_{ijk} which is defined to be (i) 0 for $\{i,j,k\}$ are not unique, (ii) +1 for $\{i,j,k\}$ is an even (clockwise) permutation, and (iii) -1 for $\{i,j,k\}$ is an odd (counter-clockwise) permutation. This expansion is verified for the case N=3 using this definition of the alternating symbol to expand the 3! = 6 triples as detailed in the slide. It is easily verified that the alternating symbol may be written as $\varepsilon_{ijk} = (i-j)(j-k)(k-i)/2!$

Direct Solution of Linear Systems

6.6.1 N-dimensional Determinants

N-dimensional Determinants

- **Expand by** *col.* α

$$|\mathbf{A}| = \sum_{k=1}^{N} a_{k\alpha} Cof(a_{k\alpha}) \qquad Cof(a_{k\alpha}) \equiv (-1)^{k+\alpha} M_{k\alpha}$$

- **Anti-symmetric Product Definition**

$$|\mathbf{A}| = \sum_{k_1 k_2 \cdots k_N = 1}^{N} \varepsilon_{k_1 k_2 \cdots k_N} \cdot a_{k_1(1)} \cdot a_{k_2(2)} \cdots a_{k_N(N)} \quad \varepsilon_{k_1 k_2 \cdots k_N} = \begin{cases} 0 & \text{for } \{k_1 k_2 \cdots k_N\} \text{ not distinct} \\ +1 & \text{for } \{k_1 k_2 \cdots k_N\} \text{ even perm.} \\ -1 & \text{for } \{k_1 k_2 \cdots k_N\} \text{ odd perm.} \end{cases}$$

- **Geometric Interpretation:** *Area, Volume, Hypervol.*

Pseudo-Determinant

$$\vec{a} \times \vec{b} = Area \cdot \hat{k} = \begin{vmatrix} \hat{i} & \hat{j} & \hat{k} \\ a_1 & a_2 & 0 \\ b_1 & b_2 & 0 \end{vmatrix} = \det \begin{vmatrix} a_1 & a_2 \\ b_1 & b_2 \end{vmatrix} \cdot \hat{k}$$

Volume as Determinant

$$(\vec{a} \times \vec{b}) \bullet \vec{c} = volume = \det \begin{bmatrix} a_1 & a_2 & a_3 \\ b_1 & b_2 & b_3 \\ c_1 & c_2 & c_3 \end{bmatrix}$$

$$A(\hat{n} \bullet \vec{c}) = A \cdot h$$

(a × b) = (Area) n̂

Cross Product General Case

Triple Product General Case

The geometric interpretation of the determinant in 3-dimensions is shown to be the volume of the parallelepiped defined by the three vectors **a, b, c** taken as either rows or columns of a matrix whose determinant we are calculating. Simple vector analysis defines area as the cross product

$$\mathbf{a} \times \mathbf{b} = |\mathbf{a}||\mathbf{b}|\sin(a,b)\,\mathbf{n} = \text{``Area''}\,\mathbf{n}$$

and the vector **n** is a unit vector perpendicular to the plane defined by the two vectors **a** & **b** according to the "right-hand-rule" It is also shown to be equivalent to the evaluation of the pseudo-determinant of the 3^d matrix having unit vectors $\hat{i}, \hat{j}, \hat{k},$ in the first row. Expanding by the first row gives components of the area along the $\hat{i}, \hat{j}, \hat{k},$-directions. For the simple case shown on the slide with $a_3 = b_3 = 0$ there is only an area component along the $\hat{k}$-direction; in the general case all three areal components exist. Similarly, the 3^d volume of the parallelepiped defined by the three vectors **a, b, c** can be obtained as the scalar product of (**a** x **b**) with a third vector **c** as follows

$$\text{Volume} = (\mathbf{a} \times \mathbf{b}) \cdot \mathbf{c} = \text{``Area''} * (\mathbf{n} \cdot \mathbf{c}) = \text{``Area''} * \text{``Height''}$$

These ideas are easily extended to hyper-volumes in higher dimensions N>3 by summing the products of N-dimensional vectors { $\mathbf{a}_1, \mathbf{a}_2, \mathbf{a}_3, ..., \mathbf{a}_N$ } using the N-dimensional alternating symbol $\varepsilon_{i1, i1, , iN}$ in anti-symmetric product definition of a determinant given on Slide#6-26.

6.6.2 Determinants - Some Properties

Determinants - Some Properties

- **Volume Interpretation makes obvious:**
  - Vol. = $\det \mathbf{A} \leq \|\mathbf{a}\| \cdot \|\mathbf{b}\| \cdot \|\mathbf{c}\| = \sqrt{a_1^2 + a_2^2 + a_3^2} \cdot \sqrt{b_1^2 + b_2^2 + b_3^2} \cdot \sqrt{c_1^2 + c_2^2 + c_3^2}$
  - Equality holds if **a, b, c** are orthogonal
  - Zero volume ➔ detA=0; this occurs whenever
    - **c=0,** or
    - **c** is a linear comb. of **a** & **b** (**c** in plane of **a** & **b**)
  - Generalizes to n-dimensions and hyper volumes
- **Triangular Matrices Expand by** *col. Minors yields*

$\det \mathbf{U} = u_{11} \cdot u_{22} \cdots u_{NN}$

$\det \mathbf{L} = l_{11} \cdot l_{22} \cdots l_{NN}$

$\det \mathbf{U}_1 = 1 \cdot 1 \cdots 1 = 1$

$$L = \begin{bmatrix} l_{11} & 0 & 0 & 0 & 0 \\ l_{21} & l_{22} & 0 & 0 & 0 \\ l_{31} & l_{32} & l_{33} & & 0 \\ \vdots & \vdots & \cdots & l_{N-1,N-1} & 0 \\ l_{N1} & l_{N2} & \cdots & l_{N,N-1} & l_{NN} \end{bmatrix}$$

$\det L = l_{11} \cdot M_{11}$
$M_{11} = l_{22} \cdot M_{22}$
$M_{22} = l_{33} \cdot M_{33}$
$\vdots$
$M_{N-1,N-1} = l_{NN} \cdot \underbrace{M_{NN}}_{=1}$

The geometric interpretation of determinants as volume (or more properly "coordinate volume") leads to some results that might otherwise not be as intuitive. For example, the fact that the determinant is equal to the volume of the *parallelepiped* constructed from row (or column) vectors **a**, **b**, and **c** allows us to conclude that the determinant is always less than or equal to the product of their magnitudes as expressed in the top equation on the slide.

The equality holds when the three vectors **a**, **b**, and **c** are orthogonal in which case the volume of the parallelepiped is exactly equal to the product of their magnitudes; moreover, the determinant vanishes if the three vectors are co-planar corresponding to zero volume of the parallelepiped. The latter circumstance corresponds to the statement that "if one row (or column) of a matrix is a linear combination of the other two (coplanar vectors), then its determinant is zero". These concepts are easily extended to higher dimensions in an obvious manner, even though we lack a clear mental picture of hyper-volumes.

The determinants of upper and lower triangular matrices as well as a "unit" triangular matrix with "ones' along its diagonal can all be computed without much work because of their special structure. Consider the determinant of the lower triangular matrix shown on the slide; we may expand this matrix by minors (co-factors) using any convenient row or column. If we judiciously choose to expand by the first row which contains only one non-zero element, then the expansion reduces to a single term $\det \mathbf{L} = l_{11} * \det(\mathbf{M}_{11})$. The minor $\mathbf{M}_{11}$ also has a 1st row with only a single non-zero element l_{22} and accordingly its determinant is $\det(\mathbf{M}_{11}) = l_{22} * \det(\mathbf{M}_{22})$. Continuing with this process along the "stepped path" in the figure, until we reach the last element l_{nn} we find that the determinant of the matrix L is equal to the product of its diagonal terms, viz., $\det \mathbf{L} = l_{11}\, l_{22} \ldots l_{nn}$.

Clearly the same argument works for the upper matrix and we have $\det \mathbf{U} = u_{11}\, u_{22} \ldots u_{nn}$ and for the "unit" triangular matrices $\mathbf{L}_1$ or $\mathbf{U}_1$ with ones along their diagonals we have $\det \mathbf{L}_1 = \det \mathbf{U}_1 = 1$.

6.7 Other Matrix Factorizations

Other Matrix Factorizations

- **General Procedure**

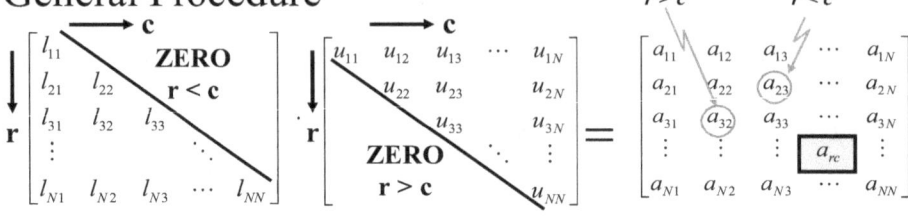

- N^2 unique a's and $(N^2 + N)$ l's & u's → N free conditions
- Write down matrix product for general element a_{rc}

$r \geq c$ "a_{32}" $a_{rc} = [l_{r(1)}u_{(1)c} + l_{r(2)}u_{(2)c} + \cdots + l_{r(c-1)}u_{(c-1)c}] + l_{r(c)}u_{(c)c} + l_{r(c+1)}\underbrace{u_{(c+1)c}}_{=0}$

$$a_{rc} = \sum_{\alpha=1}^{c-1} l_{r(\alpha)} u_{(\alpha)c} + l_{r(c)} u_{(c)c} \quad (1)$$

$r \leq c$ "a_{23}" $a_{rc} = [l_{r(1)}u_{(1)c} + l_{r(2)}u_{(2)c} + \cdots + l_{r(r-1)}u_{(r-1)c}] + l_{r(r)}u_{(r)c} + \underbrace{l_{r(r+1)}}_{=0} u_{(r+1)c}$

$$a_{rc} = \sum_{\alpha=1}^{r-1} l_{r(\alpha)} u_{(\alpha)c} + l_{r(r)} u_{(r)c} \quad (2)$$

- **CroUt, DooLittle, CholeSki, LDU Factorizations**

We found that the Gaussian Elimination procedure creates an upper triangular matrix **U** and also produces a matrix of row multipliers conveniently denoted l_{ij} which form the elements of a "unit" lower triangular matrix $\mathbf{L}_1$. The surprising fact is that the product of these two matrices $\mathbf{L}_1 \mathbf{U}$ yields the original matrix **A** and thus constitutes the factorization $\mathbf{A} = \mathbf{L}_1 \mathbf{U}$ which turned out to be very useful. Here we show how to obtain four other factorizations of the Matrix of **A**, namely Crout, Doolittle, Choleski, and **LDU**.

The general procedure is to multiply a lower triangular matrix **L** (elements l_{rc}) times an upper triangular matrix **U** (elements u_{rc}) to form the matrix **A** (elements a_{rc}) as depicted in the figure. Note that the elements of **L** are zero for r<c while those for **U** are zero for r>c. The number of distinct elements in each triangular matrix is $N^2/2 + N/2$ so the total number of unique elements in **L** and **U** combined is $N^2 + N$. However, we only have N^2 equations (one for each element of **A**) to solve for the $(N^2 + N)$ unknown elements of **L** and **U**; hence in order to get a unique solution, we must choose N of these values and these choices lead to several different factorizations.

Concentrating on a general "row/column" element a_{rc} of the **A** matrix the results depend upon whether the row or column is larger:

Case (i) $r \geq c$ a_{rc} is equal to the sum of products $l_{r\alpha} * u_{\alpha c}$ which ends when the inner index $\alpha = c$ because the very next term $\alpha = c + 1$ uses $u_{c+1, c}$ which is zero because "row index" =c+1 > c ="col index" is below the diagonal. Note, that the last term in the sum is actually taken out of the sum so the remaining sum ends $\alpha = c - 1$ as shown in the upper boxed equation.

Case (ii) $r \delta c$ a_{rc} is equal to the sum of products $l_{r\alpha} * u_{\alpha c}$ which ends when the inner index $\alpha = r$ because the very next term $\alpha = r+1$ uses $l_{r, r+1}$ which is zero because "row index" =r < r+1= "col index" is above the diagonal. Note, that again the last term in the sum is actually taken out of the sum so the remaining sum ends $\alpha = r - 1$ as shown in the lower boxed equation.

Note that separating the last terms out of the sums in the boxed equations facilitates the solution for l_{rc} in Eq.(1) and for u_{rc} in Eq.(2).

Direct Solution of Linear Systems

6.7.1 DooLittle & CroUt Factorizations

DooLittle & CroUt Factorizations

- **DooLittle**: Choose N conditions $l_{rr} = 1;\ r = 1,\ldots,N$
  - Solve Eq.(2) for u_{rc} & Eq.(1) for l_{rc}

$$u_{rc} = a_{rc} - \sum_{\alpha=1}^{r-1} l_{r(\alpha)} u_{(\alpha)c} \quad (2') \quad r \le c$$

$$l_{rc} = \frac{1}{u_{cc}} \left[a_{rc} - \sum_{\alpha=1}^{c-1} l_{r(\alpha)} u_{(\alpha)c} \right] \quad (1') \quad r > c$$

- **CroUt**: Choose N conditions $u_{cc} = 1;\ c = 1,\ldots,N$

$$l_{rc} = a_{rc} - \sum_{\alpha=1}^{c-1} l_{r(\alpha)} u_{(\alpha)c} \quad (1'') \quad r \ge c$$

$$u_{rc} = \frac{1}{l_{rr}} \left[a_{rc} - \sum_{\alpha=1}^{r-1} l_{r(\alpha)} u_{(\alpha)c} \right] \quad (2'') \quad r < c$$

- Row/column order of computation is important!!

The difference between the DooLittle and CroUt factorizations is in using the N free conditions to make either **L** or **U** a unit triangular matrix respectively.

The DooLittle factorization formulas result from setting the N diagonal elements $l_{rr} = 1$ and first using Eq.(2) to solve (without division) for u_{rc} and then using Eq.(1) to solve for l_{rc} with the diagonal divisor u_{cc}. It is key to solve this pair of equations in a specific order so that the needed terms are available at each step of the procedure as illustrated in the top sequence of figures.

Starting with the top equation setting r=1, the sum collapses and we simply have for "row$_1$ of **U**" $u_{1c} = a_{1c}$ for c=1,2,...,N as indicated by the grey row across the top of **U** in the first figure of the sequence. Next we set c = 1 in the bottom equation and again the sum collapses and we simply have for "col$_1$ of **L**" $l_{r1} = a_{r1}/u_{rr}$ for r > c (R=2,3,...,N) which (together with l_{11}=1) gives the grey "col$_1$ of **L**" as indicated by the 2nd figure in the sequence. We now have all the terms needed to solve the top equation with r=2 for "row$_2$ of **U**" $u_{2c} = a_{2c} - l_{21} u_{1c}$ for c=2,3...N as shown in the 3rd figure in the sequence. We now have all the terms needed to solve the bottom equation with c=2 for "col$_2$ of **L**" $l_{r2} = [a_{r2} - l_{r2} u_{2c}] / u_{rr}$ for r=2,3...N as shown in the 4th figure in the sequence. The procedure continues alternating between the two equations computing rows of **U** and columns of **L** until all elements in **U** and **L** have been computed.

The Crout factorization formulas result from setting the N diagonal elements $u_{rr} = 1$ and first using Eq.(1) to solve (without division) for l_{rc} and then using Eq.(2) to solve for u_{rc} with the diagonal divisor l_{rr}. Now the order of solution is the top equation for "col$_1$ of **L**", then the bottom equation for "row$_1$ of **U**", *etc.*, until all elements in **U** and **L** have been computed.

Direct Solution of Linear Systems

6.7.2 CholeSki Factorization

CholeSki Factorization

- **CholeSki :** $A = LU = L\,L^T = U^T U$
  1. Pos. Def. $Q = x^T A x > 0$, (arb. x)
  2. Sym. $L^T = U \rightarrow l_{ij} = u_{ji}$
  3. N cond. $l_{rr} = u_{rr};\ r = 1,\ldots,N$
  - Solve Eq.(1) for l_{rc}
  - Separate into diag (r = c) & off-diag (r > c)

$$l_{rc} = \frac{1}{u_{cc}}\left[a_{rc} - \sum_{\alpha=1}^{c-1} l_{r(\alpha)}\underbrace{u_{(\alpha)c}}_{=l_{c(\alpha)}}\right]_{=l_{cc}}$$

$$= \frac{1}{l_{cc}}\left[a_{rc} - \sum_{\alpha=1}^{c-1} l_{r(\alpha)} l_{c(\alpha)}\right] \quad (1''') \quad r \geq c$$

$$l_{rr} = \sqrt{a_{rr} - \sum_{\alpha=1}^{r-1} l_{r\alpha}^2} \quad r = c$$

$$l_{rc} = \frac{1}{l_{cc}}\left[a_{rc} - \sum_{\alpha=1}^{c-1} l_{r\alpha} l_{c\alpha}\right] \quad r > c > \alpha$$

Order:
1. $r = 1:\ l_{11} = \sqrt{a_{11}}$
3. $r = 2:\ l_{22} = \sqrt{a_{22} - l_{21}^2}$
5. $r = 3:\ l_{33} = \sqrt{a_{33} - (l_{31}^2 + l_{32}^2)}$
2. $r = 2, c = 1:\ l_{21} = \frac{1}{l_{11}}[a_{21}]$
4.
  - $r = 3, c = 1:\ l_{31} = \frac{1}{l_{11}}[a_{31}]$
  - $c = 2:\ l_{32} = \frac{1}{l_{22}}[a_{32} - l_{31} l_{21}]$

The CholeSki factorization requires a Positive Definite & Symmetric matrix; this allows the **A** to be expressed as the product of a lower triangular **L** and upper triangular matrix **U** which are transposes of one another $U = L^T$ and hence written in one of two symmetric forms as follows: $A = LU = LL^T = U^T U$. Using the last form of the factorization, the requirement for a Positive Definite quadratic form $Q = x^T A x$ becomes manifest $Q = x^T U^T U x = (Ux)^T U x = y^T y > 0$, where we have used the properties of the transpose to evaluate and inner product to find a positive sum of squares for all positive norm vectors $y = Ux$.

The N free conditions are used to make the diagonal elements of U and L equal, viz., $l_{rr} = u_{rr}$ and the symmetry condition $L = U^T$ requires the (i,k)-element of L to equal the (k,i)-element of L: $l_{ik} = u_{ki}$. Because of the symmetry, we can choose either of the boxed equations of Slide#6-29; choosing Eq.(1) for definiteness, and solving for l_{rc} we obtain the 1st equation on the slide which is valid for $r \geq c$.

Upon setting r=c in that equation we obtain the upper a square root expression on the bottom left for the diagonal elements l_{rr} (= u_{rr}); for $r \neq c$ we obtain the original expression repeated for convenience when discussing the order of solution. The order of solution is indicated by the numbered order sequence and associated figure to the right of this pair of equations for the diagonal and off-diagonal elements of L.

Step#1 used the top equation for r=1 to compute $l_{11} = (a_{11})^{\frac{1}{2}}$; **step#2** uses the bottom equation with r=2, c=1 to compute $l_{21} = (1/l_{11})[a_{21}]$; **step#3** is then back to top equation with r=2 to compute the next diagonal element $l_{22} = (a_{22} - l_{21}^2)^{\frac{1}{2}}$; **step#4** uses the bottom equation again with r=4, c=1,2 to compute $l_{31} = (1/l_{11})[a_{31}]$ and $l_{32} = (1/l_{22})[a_{32} - l_{31} l_{21}]$; etc. .

The figure shows the pattern, first computing a diagonal element, then dropping down one row and computing elements across it until the next diagonal element is reached and computed. The procedure continues alternating between the two equations computing diagonal terms using the top equation and off-diagonal terms using the bottom equation until all elements of **L** have been computed. Thus **A** is written in terms of **L** and its transpose L^T or in terms of **U** and its transpose U^T which are now both known since they are transposes of one another.

Direct Solution of Linear Systems

6.7.3 LDU Factorization

LDU Factorization

- **LDU** : $A = L_1 DU_1 = \boxed{L_1 D L_1^T} = U_1^T DU_1$

1. A Pos. Def. $Q = x^T A x > 0$; arb. x
2. Sym. $L_1^T = U_1$ → $l_{ij} = u_{ji}$; D is diagonal $d_{ii} > 0$ $\begin{pmatrix}\text{Req'd for positive}\\ \text{definite matrix}\end{pmatrix}$
3. N cond. $l_{rr} = u_{rr} = 1$; $r = 1,\ldots,N$

   – Let $\tilde{U} = DU_1$ → $\tilde{u}_{rc} \equiv (DU_1)_{rc} = \sum_{\alpha=1}^{N}(d_{rr}\delta_{r\alpha})\cdot u_{\alpha c} = d_{rr}u_{rc} = d_{rr}l_{cr}$

   – Thus $L_1(DU_1) = L_1\tilde{U}$ and can apply Eq.(1) with $\tilde{U}$

   $$a_{rc} = \sum_{\alpha=1}^{c-1} l_{r\alpha}\tilde{u}_{\alpha c} + l_{rc}\tilde{u}_{cc} = \sum_{\alpha=1}^{c-1} l_{r\alpha}\underbrace{\tilde{u}_{\alpha c}}_{=d_{\alpha\alpha}l_{c\alpha}} + l_{rc}\underbrace{\tilde{u}_{cc}}_{=d_{cc}l_{cc}} \quad (1)$$

   – Noting that $l_{cc} = 1$, solve for l_{rc}, and set r = c & then r > c

Solve (1) for d_{cc}

$$d_{cc} = \left[a_{cc} - \sum_{\alpha=1}^{c-1} l_{c\alpha}^2 \cdot d_{\alpha\alpha}\right] \quad r = c$$

Solve (1) for l_{rc}

$$l_{rc} = \frac{1}{d_{cc}}\left[a_{rc} - \sum_{\alpha=1}^{c-1} l_{r\alpha}l_{c\alpha}\cdot d_{\alpha\alpha}\right] \quad r > c > \alpha$$

Order

$$\begin{cases}
1 & c=1: d_{11} = a_{11} \\
3 & c=2: d_{22} = [a_{22} - l_{21}^2 \cdot d_{11}] \\
5 & c=3: d_{33} = [a_{33} - (l_{31}^2 \cdot d_{11} + l_{32}^2 \cdot d_{22})]
\end{cases}$$

$$\begin{cases}
2 & r=2, c=1 \quad l_{21} = a_{21}/d_{11} \\
4 & \begin{cases} r=3, c=1 \quad l_{31} = a_{31}/d_{11} \\ c=2 \quad l_{32} = (a_{32} - l_{31}l_{21}d_{11})/d_{22}\end{cases}
\end{cases}$$

$$\begin{matrix} d_{11} & & & & \\ l_{21} & d_{22} & & & \\ l_{31} & l_{32} & d_{33} & & \\ \vdots & & & \ddots & \\ l_{N1} & l_{N2} & l_{N3} & \cdots & d_{NN}\end{matrix}$$

The LDU factorization also requires a Positive Definite & Symmetric matrix; this allows the **A** to be expressed as the product of a lower triangular L_1, a diagonal **D**, and an upper triangular matrix U_1 ; the lower and upper triangular are are transposes of one another $U_1 = L_1^T$ and hence the result is written in one of two symmetric forms as follows: $A = L_1 DU_1 = L_1 DL_1^T = U_1^T DU_1$. Using the last form of the factorization, the requirement for a Positive Definite quadratic form $Q = x^T A x$ becomes manifest as $Q = x^T U_1^T DU_1 x = (D^{1/2}U_1 x)^T D^{1/2}U_1 x = y^T y > 0$ since $d_{ii} > 0$, where $D^{1/2}$ is the diagonal matrix with square root of the diagonals. We have used the properties of the transpose to evaluate and inner product to find a positive sum of squares for all positive norm vectors $y = U_1 x$. The N free conditions are used to make the diagonal elements of both U and L equal to 1, viz., $l_{rr} = u_{rr} = 1$ and the symmetry condition $L = U^T$ requires the (i,k)-element of L to equal the (k,i)-element of L: $l_{ik} = u_{ki}$. Defining U_1-tilde = DU_1 casts **A** into the form $A = L_1 DU_1 = L_1 (U_1\text{-tilde})$ and noting that the elements of $(U_1\text{-tilde})_{rc} = d_{rr} u_{rc} = d_{rr} l_{cr}$ Eq.(1) is rewritten as given in the center equation for a_{rc}. Setting r=c and noting that the diagonal elements $l_{cc} = 1$ we obtain the 1st equation for the diagonal elements of d on the left bottom. For r>c we obtain the second equation for the off diagonal elements l_{rc}. Note the elements of $(U_1\text{-tilde})_{rc}$ are written as the matrix product of **D** and U_1

$Sum_\alpha\{(D)_{r\alpha}(U_1)_{\alpha c}\} = \Sigma_\alpha\{d_{rr}\delta_{r\alpha}u_{\alpha c}\} = d_{rr} u_{rc} = d_{rr} l_{cr}$.

The solution procedure for these two equations is nearly identical to that for Choleski except now the diagonal equation elements are for a separate matrix **D** ; the diagonal elements for L_1 are "1" by definition. The off-diagonal steps solve for the off-diagonal elements of L_1. Note that in contrast with Choleski, there are no square roots needed in this solution; this is a direct consequence of introducing the auxiliary diagonal matrix **D** in the $L_1 D L_1^T$ factorization of **A**.

For the purposes of showing the computational flow, we have placed the unknown diagonal matrix elements d_{kk} where the (known) diagonal elements of the L_1 matrix would ordinarily be. The figure then shows the computational sequence starting with the computation of the 1st diagonal element of **D**, then dropping down one row and computing elements across the L_1 matrix until the 2nd diagonal of **D** is reached and computed. The procedure continues alternating between the two equations computing diagonal terms of **D** using the top equation and off-diagonal terms of L_1 using the bottom equation.

6.7.4 Diagonal Dominant Matrix

Diagonal Dominant Matrix

- Strict Diagonal Dominance (SDD): $\left|diag_{kk}\right| > \sum_{k}\left|row_{k}\right|$

- Example:
$$A = \begin{bmatrix} (7) & 2 & 0 \\ 3 & (5) & -1 \\ 0 & 5 & (-6) \end{bmatrix} \begin{array}{l} \leftarrow |(7)| > |2| + |0| \\ \leftarrow |(5)| > |3| + |-1| \\ \leftarrow |(-6)| > |0| + |5| \end{array} \right\} \text{A diag. dom.}$$

$|5| \underset{NO!!}{\geq} |2| + |5|$ A^T not diag dom.

- **Theorem: If A is Strictly Diag. Dominant, Then**
  i. A is nonsingular; A^{-1} exists; $\det(A) \neq 0$
  ii. $Ax = b$ solved by Gaussian Elim. **w/o pivoting**
  iii. Solution is stable to RO error growth

The magnitude of the diagonal elements of the **A** matrix play a key role in controlling the ***round off error*** in a numerical solution to the basic matrix equation $Ax = b$ and it was for this reason that we introduced the idea of "pivoting" (or swapping rows). There is a class of matrices which do not require any pivoting as stated in the Theorem on the slide.

A matrix is said to be Strictly Diagonal Dominant (SDD) if for each row the magnitude of its diagonal element is greater than the sum of magnitudes of the remaining elements in that row. Thus, in the example matrix **A**, we satisfy the criterion row-by-row as follows: row#1 $|7| > |2| + |0|$, row#2 $|5| > |3| + |-1|$ and row#3 $|-6| > |0| + |5|$. Note that although **A** is SDD its transpose A^T is not necessarily SDD; in our example the transpose is not SDD. This is easily verified by summing down the **columns of A** (which is equivalent to summing across the **rows of A^T**); it turns out that the middle column does not satisfy the inequality $|5| > |2| + |5|$ so A^T is not SDD.

The Theorem states that if **A** is Strictly Diagonally Dominant, then (i) **A** is non-singular, (ii) $Ax=b$ does not require pivoting, and (iii) The solution is stable to RO error growth. Thus a little work up front spent on inspecting the matrix for the SDD property saves us the computational cost involved in pivoting during the Gaussian Elimination procedure.

6.7.5 Positive Definite Matrix

Positive Definite Matrix

- **A positive definite**
  - Quadratic form: $Q = x^T A x > 0$; arb. x
  - A is symmetric: $A = A^T \Longrightarrow a_{ij} = a_{ji}$

$$A = \begin{bmatrix} a_{11} & a_{12} & a_{13} \\ a_{21} & a_{22} & a_{23} \\ a_{31} & a_{32} & a_{33} \end{bmatrix}$$ *Determinants of all leading principal submatricies > 0*

- Example: Arb $[x,y,z]$, Quadratic form is a sum of squares >0

1) Leading principal submatrix determinants all positive

$\det[2] > 0 \quad \det\begin{bmatrix} 2 & -1 \\ -1 & 2 \end{bmatrix} = 4-1 = 3 > 0$

$\det\begin{bmatrix} 2 & -1 & 0 \\ -1 & 2 & -1 \\ 0 & -1 & 2 \end{bmatrix} = (2)\cdot\det\begin{bmatrix} 2 & -1 \\ -1 & 2 \end{bmatrix} - (-1)\cdot\det\begin{bmatrix} -1 & -1 \\ 0 & 2 \end{bmatrix} = 4 > 0$

2) Direct algebraic verification for arbitrary 3-vector

$Q = \begin{bmatrix} x & y & z \end{bmatrix} \begin{bmatrix} 2 & -1 & 0 \\ -1 & 2 & -1 \\ 0 & -1 & 2 \end{bmatrix} \cdot \begin{bmatrix} x \\ y \\ z \end{bmatrix}$

$Q = 2x^2 + 2y^2 + 2z^2 - 2xy - 2yz$
$= \underbrace{(x^2 - 2xy + y^2)}_{=(x-y)^2} + \underbrace{(y^2 - 2yz + z^2)}_{=(y-z)^2} + x^2 + z^2$

$\therefore Q > 0 \quad \text{pos. def.}$

- **Theorem #1**: *If A is pos. def. then*
  i. A is nonsingular
  ii. All diag. El. Are positive $a_{ii} > 0$
  iii. Max (El.) = Max (diag. El.)
  iv. Sq(off-diag) < diag prods. $a_{ij}^2 < |a_{ii}|\cdot|a_{jj}|$

- **Theorem #2**: *If A is pos. def. then*
  i. Gaussian Elim. $Ax = b$ w/o pivoting
  ii. A pos def $\Leftrightarrow A = L_1 D L_1^T$ D pos. diagonal; L_1 has "1"s on diagonal
  iii. A pos def $\Leftrightarrow A = LL^T$ l_{ii} non zero (pos or neg)

We have seen that the Choleski and **LDU** Matrix factorizations required the matrix **A** to be *positive definite and symmetric*, which means that the quadratic form $Q = x^T A x > 0$ for *arbitrary vectors* **x** and that $A = A^T$ (or $a_{rc} = a_{cr}$). Symmetry of a matrix **A** is easily verified; however positive definiteness of **A** is much more difficult since it requires the quadratic form to be greater than zero for any and all vectors **x**. An alternate method to is equally difficult as it requires the determinants of all leading principal submatrices to be positive.

The positive definiteness of the example symmetric matrix **A** given on this slide is easily verified by computing the determinants or directly by completing the squares of the multi-variable polynomial thereby yielding a sum of squares of various combinations of the components of **x**=[x y z]. However, this is not typical and usually a lot of work is needed to "vet" a given matrix as positive definite. However, once positive definiteness is verified, there are two powerful theorems that make solution of the linear system more straight forward and computationally efficient.

Theorem #1 states that a positive definite matrix **A** has the following additional properties: (i) it is non-singular, (ii) all its diagonal elements are positive $a_{ii} > 0$, (iii) its maximum element is on the diagonal, (iv) the square of any off-diagonal element less than the magnitude product the diagonal elements corresponding to its row and column index $a_{rc}^2 < |a_{rr}||a_{cc}|$

Theorem #2 states that a positive definite matrix **A** (i) does not require pivoting in Gaussian Elimination, (ii) Positive definiteness is equivalent to **A** has **LDU** factorization with $d_{ii} > 0$, and (iii) Positive definiteness is equivalent a Choleski factorization of **A** with diagonal elements l_{ii} positive or negative (just non-zero)

7 Iterative Solution of Linear Systems

n x n Linear Systems
Iterative Solutions

Tools of the Trade
Iterative Methods
Jacobi, Gauss-Seidel, Relaxation (SOR)
Condition Number & Iterative Refinement of Direct Solution

We have discussed methods for finding explicit deterministic solutions to $\mathbf{Ax} = \mathbf{b}$ by algebraic manipulation of the system of equations and now we consider how iterative solutions can be used to advantage when the number of equations becomes large.

In root finding we stopped the iteration sequence by specifying a tolerance on the difference between successive terms and tested for convergence by simply making a scalar comparison between two numbers. On the other hand, the iterative solution to the system of equations is a vector sequence $\{x^{(0)}, x^{(1)}, ..., x^{(k)}\}$ and thus we need to develop ideas about the "closeness" of two vectors $x^{(k)}$ and $x^{(k-1)}$ rather than just a scalar comparison. Thus, we need to develop the concept of vector norms and other "tools of the trade" before we can develop the iterative methods. Of the many methods available, we restrict ourselves to the few fundamental ones: Jacobi, Gauss-Seidel, and Relaxation methods.

The Condition Number of a matrix $K(\mathbf{A})$ determines how "ill-conditioned" and thus how difficult the Gaussian elimination solution to the matrix equation $\mathbf{b} = \mathbf{Ax}$ will be. Applying the matrix $\mathbf{A}$ to estimated solution $\mathbf{x}_{hat}$ yields the estimated data value $\mathbf{b}_{hat} = \mathbf{A}\,\mathbf{x}_{hat}$. One may define state and data residual vectors respectively as $\Delta \mathbf{x} = \mathbf{x} - \mathbf{x}_{hat}$ and $\Delta \mathbf{b} = \mathbf{b} - \mathbf{b}_{hat} = \mathbf{b} - \mathbf{A}\,\mathbf{x}_{hat}$. The condition number $K(\mathbf{A})$ is the "force multiplier" which when applied to the fractional change $\|\Delta \mathbf{b}\| / \|\mathbf{b}\|$ in the data residual yields the fractional change in the state vector $\|\Delta \mathbf{x}\| / \|\mathbf{x}\|$. A "large" $K(\mathbf{A})$ value signals when a small fractional change in data residuals $\|\Delta \mathbf{b}\| / \|\mathbf{b}\|$ is not sufficient to assure a small $\|\Delta \mathbf{x}\| / \|\mathbf{x}\|$ in the state vector. For this reason, we must always test for a large value of $K(\mathbf{A})$, because it signals large errors in the state vector and the need for an iterative refinement technique in order to obtain an accurate solution.

Iterative Solution of Linear Systems

7.1 Vector Norms

Vector Norms

- Vector iterates need stopping procedure
- Scalar magnitude to test against a tolerance
- p-norm $\|\mathbf{x}\|_p = \left(\sum_{k=1}^{N} |x_k|^p\right)^{1/p}$
  - 1-norm (column norm) $\quad p=1:\ \|\mathbf{x}\|_1 = \sum_{k=1}^{N} |x_k| \quad$ p=1
  - 2-norm (Euclidean) $\quad p=2:\ \|\mathbf{x}\|_2 = \left(\sum_{k=1}^{N} |x_k|^2\right)^{1/2} \quad$ p=2
  - max-norm (row norm) $\quad p=\infty:\ \|\mathbf{x}\|_\infty = \lim_{p \to \infty}\left(\sum_{k=1}^{N} |x_k|^p\right)^{1/p} \quad$ p=∞
    $= \max\{|x_1|, |x_2|, |x_3|, \cdots |x_N|,\}$
- Norms satisfy

(i) $\|\mathbf{x}\| \geq 0$ all $\mathbf{x} \in R^{(N)}$; $\|\mathbf{x}\| = 0$ iff $\mathbf{x} = \mathbf{0}$ Pos. Def.

(ii) $\|\alpha \mathbf{x}\| = |\alpha| \cdot \|\mathbf{x}\|$ scalar mult.

(iii) $\|\mathbf{x} + \mathbf{y}\| \leq \|\mathbf{x}\| + \|\mathbf{y}\|$ Norm(sum) ≤ Sum(norms)

Can also show that

(iv) $\|\mathbf{x} \cdot \mathbf{y}\| \leq \|\mathbf{x}\| \cdot \|\mathbf{y}\|$ Norm(prod) ≤ Prod(norms)

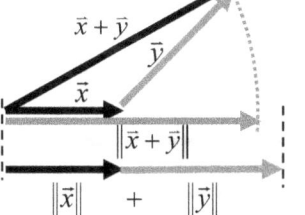

In order to have an effective stopping criterion for a vector sequence $\{x^{(0)}, x^{(1)}, ..., x^{(k)}\}$ we need to define the norm of a vector. There is an infinite family of so-called "p-norms" which are defined to be the "p^{th} root of the sum of component magnitudes raised to the pth power" for p=1,2,3,..., ∞.

The most useful norms for our purposes are the 1-norm, 2-norm, and ∞-norm (also known as the max-norm) and they are written down explicitly and illustrated in the top figure on the slide. We note that the 1-norm is just the sum of component magnitudes, the 2-norm is the familiar Euclidean square root of sum of squares norm and the max norm is an unfamiliar "limiting norm" as the integer p→ ∞.

The top figure illustrates these norms in 2-space by showing the location of the endpoints of all "unit vectors" emanating from the origin; the 2-norm unit vectors all "land" on the unit circle. The max-norm unit vectors land on the square surrounding the unit circle; the x and y coordinates both range over [-1,1]; this may seem a little strange, but all the following vectors with endpoints on that box [±1, ±1], [0, ±1], [±1, 0] have a max-norm of "1". Finally the 1-norm unit vectors all end on the inner square as shown; this is easily verified in the 1st quadrant since both x and y are positive |x|+|y| = x+y =1 means y = 1-x which is the negative one slope line starting at the point (0,1) and ending at the point (1,0). Note that all the other p-norms are between the 1-norm diamond and the max-norm square.

The reason the ∞-norm is also called the "max-norm" is because raising each component magnitude to the "p^{th} power" always makes the largest element dominant and hence in taking the pth root of the sum always "picks" the max component in the limit as p → ∞.

All norms must satisfy the three properties given on the slide which basically state that (i) the norm of a vector x must be positive definite, (ii) multiplication of each component by the scalar "a" yields "a" times the original norm, and (iii) "the norm of the sum (of two vectors) is always less than or equal to the sum of the two norms." The lower figure illustrates property (iii) showing that the equality holds when **x** and **y** are co-linear so their magnitudes simply add ; in general, the inequality holds. These three defining properties of a norm can be used to prove a 4th one, namely, (iv) "the norm of the product (of two vectors) is always less than or equal to the product of the two norms"

Iterative Solution of Linear Systems

7.1.1 Norm Inequalities

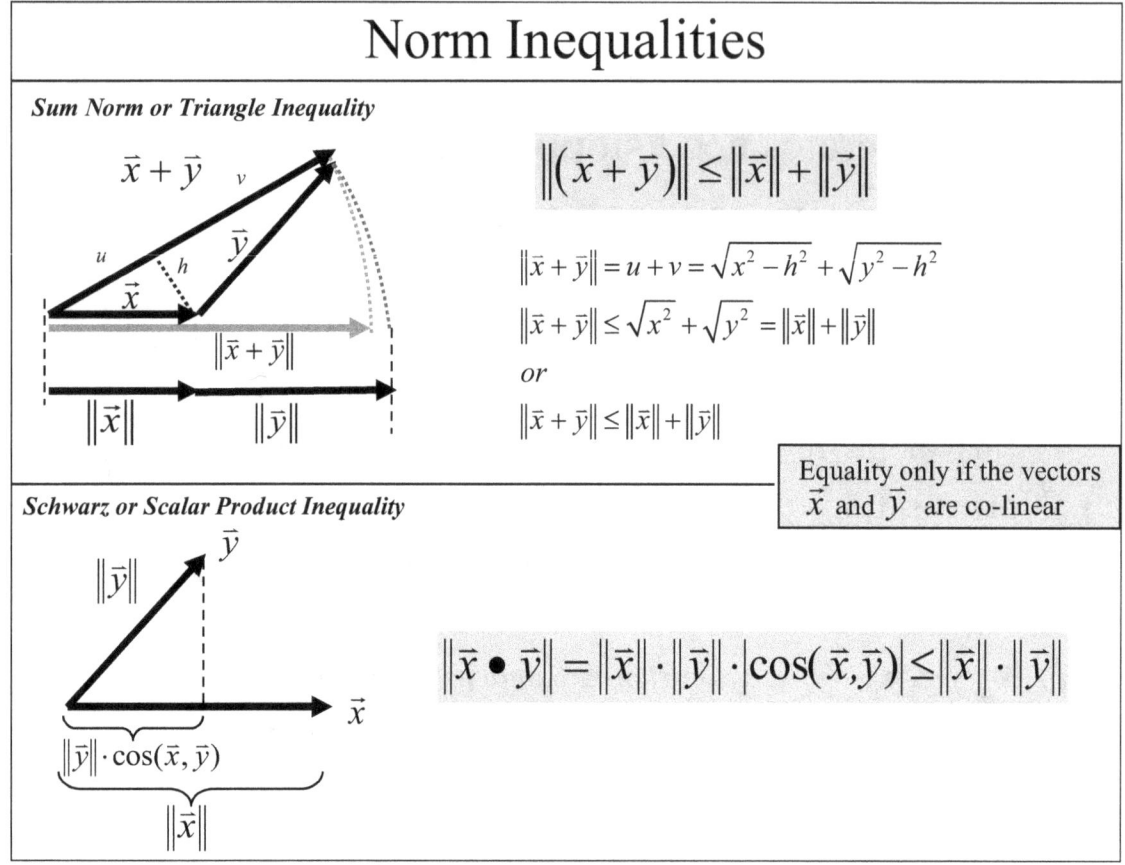

Here is a geometric and an algebraic proof of the sum norm and Schwarz inequalities stating respectively that the norm of the sum vector is less than or equal to the sum of their norms and norm of the scalar product is less than or equal to the product of the norms. Equality holds for co-linear vectors.

Sum Norm (Vector Triangle) Inequality:
For the geometric analysis the two non-co-linear vectors **x** and **y** are summed graphically from the tail of **x** to the tip of **y** to yield the new vector (**x**+**y**). A rotated version (**x**+**y**)$_{rot}$ is shown as the grey arrow lying along the x-axis and the dashed circular arc emphasizes that its *magnitude is unchanged by rotation*. If the **y** vector is also rotated onto the x-axis (blue circular arc) and then has its magnitude added to the magnitude of **x** we obtain the two-piece black arrow representing the sum ||**x**||+||**y**|| and it is seen to be larger than the grey vector with magnitude || **x**+**y** ||.

In the algebraic proof we consider the vector triangle and simply drop the perpendicular "h" onto sum vector side forming two small right triangles. Then observe that the magnitude ||**x**+**y**|| = u +v and further that u = $(x^2-h^2)^{1/2}$ and v= $(y^2-h^2)^{1/2}$ to obtain top equation on the slide which immediately gives the desired inequality. Clearly the equality can only occur for $h=0$ in which case the triangle collapses to a straight line.

Schwarz product inequality: A geometric proof is given by using the definition of a scalar product as the projection of one vector onto the other or the product of ||**y**| |cos(**x**,**y**)| with ||**x**|| the magnitude of the second vector **x** . Since the cosine function has magnitude |cos(**x**,**y**)| ≤1 the result follows.

Iterative Solution of Linear Systems

7.2 Matrix Norms

Matrix Norms

- **Natural (Induced) Norms**
  - $y = Ax$ is a vector
  - Find the worst case (**max.**) for all possible unit vectors $\bar{x}$ relative to same vector norm

Definitions:

$$\|A\|_1 = \max\{\|Ax\|_1\}_{\|x\|_1 = 1}$$

$$\|A\|_2 = \max\{\|Ax\|_2\}_{\|x\|_2 = 1}$$

$$\|A\|_\infty = \max\{\|Ax\|_\infty\}_{\|x\|_\infty = 1}$$

- **Example:**

a) row norm = max norm

$$A = \begin{bmatrix} 1 & 2 & -1 \\ 0 & 3 & -1 \\ 5 & -1 & 1 \end{bmatrix} \quad \begin{array}{l} row_1 = |1|+|2|+|-1| = 4 \\ row_2 = |0|+|3|+|-1| = 4 \\ row_3 = |5|+|-1|+|1| = 7 \end{array} \quad \|A\|_\infty = \max\{4,4,7\} = 7$$

b) col. Norm = 1-norm ↓ ↓ ↓
$\quad\quad\quad\quad\quad\quad\quad\quad col_1\ col_2\ col_3$
$\quad\quad\quad\quad\quad\quad\quad\quad\ 6\ \ \ \ 6\ \ \ \ 3 \quad \|A\|_1 = \max\{6,6,3\} = 6$

Properties:

(i) $\|A\| \geq 0$; $\|A\| = 0$ iff $A = 0_{NN}$

(ii) $\|\alpha A\| = |\alpha| \cdot \|A\|$

(iii) $\|A + B\| \leq \|A\| + \|B\|$

Can also show that

(iv) $\|A \cdot B\| \leq \|A\| \cdot \|B\|$

Since the result of multiplying a matrix **A** by a vector **x** is another vector **y**, the natural way to define the norm on a matrix is to compute the norm of the resulting output vector **y** for all possible input vectors **x**. We therefore define the Natural or Induced Norm of a matrix **A** to be the maximum output vector norm $\max\{\|y\|\}$ that results from allowing all possible unit-norm input vectors **x**.

Thus in the upper right box we define the 1-norm as $\|A\|_1 = \max\{\|Ax\|\}$ for all possible inputs **x** having a unit 1-norms $\|x\|_1=1$; similarly $\|A\|_2 = \max\{\|Ax\|\}$ for all possible inputs **x** having a unit 2-norms $\|x\|_2=1$ and $\|A\|_\infty = \max\{\|Ax\|\}$ for all possible inputs **x** having a unit ∞-norms $\|x\|_\infty=1$.

Given this definition, the same properties for vector norms now apply to matrix norms as shown in the properties box of the slide.

Thus all matrix norms must satisfy (i) the norm of a matrix $\|A\|$ must be positive definite, (ii) multiplication of each component by a scalar "a" yields "a" times the original norm, and (iii) "the norm of the sum of two matrices $\|A + B\|$ is always less than or equal to the sum of the two matrix norms $\|A\| + \|B\|$." and again these three defining properties of a norm can be used to prove a 4th one, namely, (iv) "the norm of the product of two matrices $\|AB\|$ is always less than or equal to the product of the two norms $\|A\|\ \|B\|$"

It can be shown that the **max-norm** is obtained by maximizing the **absolute row sums** and that the **1-norm** is obtained by maximizing the **absolute column sums**. The 2-norm is the most familiar and also requires the most work to compute; we shall see that it requires that we compute the eigenvalues of a related matrix $A^T A$.

The computation of the row-norm and column-norm for a simple 3 x 3 matrix is given in the example on the bottom left box of the slide and is simply a matter of summing the magnitudes of all rows or all columns and the choosing the maximum value in each case.

7.2.1 Convergence of a Vector Sequence

Convergence of a Vector Sequence

- **Distance = Norm of vect=or difference**
  - **Euclidean(2-norm):** $\|\mathbf{x}-\mathbf{y}\|_2 = \sqrt{\sum_{i=1}^{n}(\bar{x}_i - \bar{y}_i)^2}$
  - **Max (∞-norm):** $\|\mathbf{x}-\mathbf{y}\|_\infty = \max_{1 \le i \le n} |x_i - y_i|$
- **Vector Sequence:** $\{\mathbf{x}^{(k)}\}_{k=1}^{\infty} = \{x^{(1)}, x^{(2)}, \cdots x^{(N(\varepsilon))}, \cdots\}$
- **Converges:** $\lim_{n \to \infty} \mathbf{x}^{(n)} = \mathbf{x}$

  iff $\|\mathbf{x}^{(n)} - \mathbf{x}\| < \varepsilon$ for all $n \ge N(\varepsilon)$

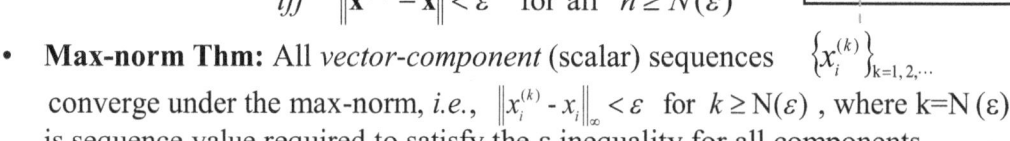

- **Max-norm Thm:** All *vector-component* (scalar) sequences $\{x_i^{(k)}\}_{k=1,2,\cdots}$ converge under the max-norm, i.e., $\|x_i^{(k)} - x_i\|_\infty < \varepsilon$ for $k \ge N(\varepsilon)$, where k=N(ε) is sequence value required to satisfy the ε inequality for all components

Proof: Letting $k > N = \max_{i=1,\cdots,n}(N_i(\varepsilon))$, clearly this k >N value insures convergence of the even the *slowest converging* component; hence *all other components* must satisfy the inequality. Since the **max-norm** chooses the largest *vector-component (which must also converge)*, we have

$$\|\bar{x}^{(k)} - \bar{x}\|_\infty = \max_{i=1,2,\cdots n} |x_i^{(k)} - x_i| < \varepsilon \text{ for all } k > N(\varepsilon)$$

Now that we have the concept of a norm for vectors and matrices we can define closeness of two vectors **x** and **y** as the norm of their difference ||**x** - **y**|| which is the familiar Euclidean distance between the two vectors for the 2-norm and the maximum (absolute) component difference for the max-norm.

For a sequence of vectors {**x**$^{(k)}$} convergence to a limit vector **x** is defined by the norm of the difference vector in a manner exactly analogous to that used for convergence of a sequence of scalars. The vector converges, if for all scalar components, there exists a single integer k= N(ε) such that the difference vector satisfies the inequality ||**x**$^{(k)}$ -**x** || < ε for all succeeding iterates k >N and for all vector components i= 1,2,...,n. The notional sketch shows the (red) limit vector **x** and a number of vectors in the converging.

No particular norm is mentioned in this definition of convergence of a vector sequence, so we are free to test convergence by specifying a specific norm; this generally does not guarantee convergence with respect to any other norm. However, there are theorems that relate convergence with respect to one norm to the convergence with respect to other norms.

The Max-Norm Theorem states that convergence of a vector sequence with respect to the max norm is equivalent to the convergence of each of its scalar component sequence $x_i^{(k)}$ to x_i. (The subscript denotes the scalar components i =1,2,3,...,n and the parenthesized superscript refers to the iterate.) The proof of the theorem simply uses the property that, although each component "i" may require a different $N_i(\varepsilon)$, there is a iteration# k=N(ε) for the "slowest converging component" (whichever that might be). This value of "k" *insures convergence of all components* for k > N; in particular the maximum component or max-norm converges as k approaches infinity.

7.3 Eigenvalues & Eigenvectors

Eigenvalues & Eigenvectors

- Need for 2-norm computation
- Vector space interpretation
- Geometrical interpretation
- Computation
- Canonical Matrix Representation
- Spectral Radius & 2-norm

The Euclidean or 2-norm of a matrix $\|A\|_2$ results from performing the matrix multiplication Ax for all vectors x whose 2-norm $\|x\|_2 = 1$, and then find the maximum component for all possible resulting vectors $y = A x$. There are an infinite number of ways to choose the components of a 2-norm unit vector; if we restrict ourselves to 2 dimensions, choosing the 1st component to be any number α in the interval [0,1], then the 2nd component of the unit vector is $(1-\alpha^2)^{1/2}$ and we have an infinite number of such vectors. Since we cannot test this infinite set of possible unit vectors $[\alpha, (1-\alpha^2)^{1/2}]^T$ the question is "how do we compute the 2-norm of the matrix A in a finite amount of time?"

The answer requires us to consider the structure of the matrix A in terms of its fundamental eigenvalues and eigenvectors. This structure is given an intuitive geometrical interpretation and the so called canonical representation of the matrix A is developed from the geometry. Next we develop a direct method for computing the eigenvectors and eigenvalues which requires that we find the roots of a characteristic polynomial whose degree is equal to the dimension of the vector x. Finally, the spectral radius $\rho(A)$ of a matrix A is defined and then the Spectral Radius Theorem relates the 2-norm $\|A\|_2$ to the spectral radius ρ of the associated product matrix $A^T A$, $viz.$, $\|A\|_2 = [\rho(A^T A)]^{1/2}$.

Iterative Solution of Linear Systems

7.3.1 Eigenvalues & Eigenvectors

Eigenvalues & Eigenvectors

$$A\vec{u}_1 = \lambda_1 \vec{u}_1$$
$$A\vec{u}_2 = \lambda_2 \vec{u}_2$$

- λ_1, λ_2 Eigenvalues = Stretch/contract along eigenvectors $\mathbf{u}_1, \mathbf{u}_2$
- Rotate to coincide with principal axes of the Ellipse

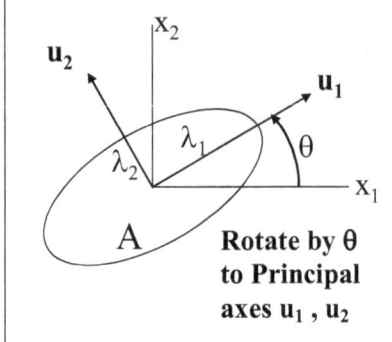

Rotate by θ to Principal axes $\mathbf{u}_1, \mathbf{u}_2$

Rewrite two equations as a single matrix equation

$$A[\vec{u}_1 : \vec{u}_2] = [\vec{u}_1 : \vec{u}_2]\begin{bmatrix} \lambda_1 & 0 \\ 0 & \lambda_2 \end{bmatrix} = \begin{bmatrix} u_{11} & u_{12} \\ u_{21} & u_{22} \end{bmatrix}\begin{bmatrix} \lambda_1 & 0 \\ 0 & \lambda_2 \end{bmatrix} = \begin{bmatrix} u_{11}\lambda_1 & u_{12}\lambda_2 \\ u_{21}\lambda_1 & u_{22}\lambda_2 \end{bmatrix}$$

$$AU = U\Lambda$$

Left multiply by U^{-1}
Solve for diagonal Eigenvalue Matrix

$$U^{-1}AU = \underbrace{U^{-1}U}_{=I} \Lambda = \Lambda = diag(\lambda_1, \lambda_2)$$

The eigenvalue-eigenvector structure of a matrix **A** is best understood geometrically as an ellipse in two 2-dimensions (ellipsoid in 3-dimensions) whose major axis is oriented at an angle θ to the x_1 axis as shown in the figure. The principal axis is along the direction of a new x_1-axis labeled u_1 in the figure and the two principal axes of the ellipse obviously lie along the rotated coordinates (u_1,u_2). For this reason, the (u_1,u_2) coordinate system is described as the principal axis system or as canonical coordinates.

The geometrical significance of the ellipse in canonical coordinates is formalized by writing down the "action" of the matrix **A** on the two unit vectors along these coordinate axes, *i.e.*, $\mathbf{u}_1$ and $\mathbf{u}_2$. The equations show that **A** acts on the vector $\mathbf{u}_1$ by stretching it by a factor of λ_1 along its coordinate direction and similarly for $\mathbf{u}_2$ by a factor λ_2 along its coordinate direction.

In contrast consider the action of the **A** matrix on a vector that is not along one of these two principal directions, such as a vector $\mathbf{e}_2$ along the x_2 coordinate axis. The action of **A** on $\mathbf{e}_2$ can be understood by breaking it up into two non-colinear components along the (u_1,u_2) axes so that $\mathbf{e}_2 = a_1 \mathbf{u}_1 + a_2 \mathbf{u}_2$. The result of the action **A** $\mathbf{e}_2$ is clearly to multiply the component along $\mathbf{u}_1$ by λ_1 and the component along $\mathbf{u}_2$ by λ_2 so the resulting vector is no longer simply stretched along its original x_2 axis direction, but is instead at some angle to the x_2 axis.

Thus the principal axes of the "A-ellipse" defines the unique directions $\mathbf{u}_1$ and $\mathbf{u}_2$ in which the action of **A** on a vector is pure stretching along the same direction; any other vector **e** can be expressed as a linear combination of $\mathbf{u}_1$ and $\mathbf{u}_2$ and its components will be scaled differently yielding a new vector that is in a completely different direction.

Iterative Solution of Linear Systems

7.3.2 Effect of Matrix A on Arbitrary Vector x

Effect of Matrix **A** on Arbitrary Vector **x**

Eigenvalue-Eigenvector Equations $\quad AU = U\Lambda \quad\quad A\mathbf{u}_k = \lambda_k \mathbf{u}_k \quad k = 1,2$

1) Eigenvectors not "Orthogonal" $\quad A = U\Lambda U^{-1} \quad\quad U = [\mathbf{u}_1 : \mathbf{u}_2]$

2) Eigenvectors "Orthogonal"
$UU^T = I \Rightarrow U^{-1} = U^T$

$$A = U\Lambda U^T = [\mathbf{u}_1 : \mathbf{u}_2]\begin{bmatrix} \lambda_1 & 0 \\ 0 & \lambda_2 \end{bmatrix}\begin{bmatrix} \mathbf{u}_1^T \\ \mathbf{u}_2^T \end{bmatrix} = [\mathbf{u}_1 : \mathbf{u}_2]\begin{bmatrix} \lambda_1 \mathbf{u}_1^T \\ \lambda_2 \mathbf{u}_2^T \end{bmatrix}$$

Orthogonal Spectral Decomposition of A
Sum of two 2 x 2 orthogonal matrices
(outer products along principal axes)

$$A = \sum_{k=1}^{2} \lambda_k \mathbf{u}_k \mathbf{u}_k^T = \lambda_1 \mathbf{u}_1 \mathbf{u}_1^T + \lambda_2 \mathbf{u}_2 \mathbf{u}_2^T$$

Write **arbitrary** 2^d vector **x** as: $\quad \mathbf{x} = x_1 \mathbf{u}_1 + x_2 \mathbf{u}_2$

A operating on **x** yields: $\quad A\mathbf{x} = A(x_1 \mathbf{u}_1 + x_2 \mathbf{u}_2) = x_1 \underbrace{A\mathbf{u}_1}_{=\lambda_1 \mathbf{u}_1} + x_2 \underbrace{A\mathbf{u}_2}_{=\lambda_2 \mathbf{u}_2}$

Output is not a simple *scaling:* $\quad A\mathbf{x} = \lambda_1 x_1 \mathbf{u}_1 + \lambda_2 x_2 \mathbf{u}_2$

Stretches by diff. amounts along each eigenvector $\quad$ (not λ times **x**)

We have found that the eigenvalue-eigenvector equations can either be written down separately for each index i=1,2 or written in the matrix form **AU=UΛ**. We can formally solve this equation for **A** by multiplying from the right by U^{-1} to obtain the general result 1) **A=UΛU^{-1}** true for any matrix **A**. Recall that U is a partitioned 2 x 2 matrix with the eigenvector $\mathbf{u}_1$ components down col#1 and eigenvector $\mathbf{u}_2$ components down col#2, so that U^{-1} is the inverse of this partitioned matrix U formed from the unit eigenvectors of A.

For the special case in which the unit eigenvectors $\mathbf{u}_1$ and $\mathbf{u}_2$ are orthogonal, the matrix U=[$\mathbf{u}_1$:$\mathbf{u}_2$] becomes orthogonal as well. By definition orthogonality means that **UU^T** = **I**, so multiplying this expression from the left by U^{-1} leads to the equivalence of the transpose and the inverse, *i.e.*, $U^T = U^{-1}$. This, in turn, allows us to write a new expression 2)**A=UΛU^T** for the case of orthogonal eigenvectors. Expanding this expression out as shown in the slide, we find that **A** can be re-expressed as an Orthogonal Spectral Decomposition which consists of the sum of two 2 x 2 matrices uniquely formed from the outer products of the eigenvectors and their associated eigenvalues, *viz.*, $\lambda_1 \mathbf{u}_1 \mathbf{u}_1^T$ and $\lambda_2 \mathbf{u}_2 \mathbf{u}_2^T$ (grey boxed equation). Note that if λ_1 =1000 and λ_2 =1, then the magnitude of the first term in the spectral expansion is dominant and thus we may neglect the 2nd term and have to good approximation **A** ≈ 1000 $\mathbf{u}_1 \mathbf{u}_1^T$. This approximation technique for a large n x n matrix **A** can be very useful when there are only a few dominant eigenvalues in the spectral decomposition. In that case, we drop all terms below a cutoff eigenvalue $\lambda_k < \lambda_{cutoff}$ and approximate **A** using the few remaining large terms. Note that for non-orthogonal basis vectors there is no such spectral decomposition and instead we have the general matrix factorization A=UAU^{-1} (see Slide# 7-10).

Now an arbitrary vector **x** may always be written as a linear combination **x** = = $x_1 \mathbf{u}_1$ +$x_2 \mathbf{u}_2$ of the orthogonal unit basis vectors {$\mathbf{u}_1$, $\mathbf{u}_2$}. Thus the result of **A** operating on **x** naturally separates into its operation on the orthogonal components of the vector **x** by simply multiplying the x_1-component by λ_1 and the x_2-component by λ_2 to yield an output vector

$$\mathbf{y} = \mathbf{A}\mathbf{x} = \mathbf{A}(x_1 \mathbf{u}_1 + x_2 \mathbf{u}_2) = x_1 (\lambda_1 \mathbf{u}_1) + x_2 (\lambda_2 \mathbf{u}_2)$$

We note that the components along $\mathbf{u}_1$ and $\mathbf{u}_2$ are scaled by different eigenvalues so the result is no longer parallel to the original input vector **x**.

7.3.3 Computing Eigenvalues/Eigenvectors

Computing Eigenvalues/Eigenvectors

- Vectors along principal axes undergo pure scaling:

$$A\bar{u} = \lambda \cdot \bar{u} = \lambda \cdot I \cdot \bar{u}$$
$$\Rightarrow (A - \lambda \cdot I)u = 0$$

- Solution requires $\boxed{\det(A - \lambda \cdot I) = 0}$

- Example: $A = \begin{bmatrix} 1 & 2 \\ 3 & 2 \end{bmatrix}$ $A - \lambda \cdot I = \begin{bmatrix} 1 & 2 \\ 3 & 2 \end{bmatrix} - \lambda \begin{bmatrix} 1 & 0 \\ 0 & 1 \end{bmatrix}$

$$\det \begin{bmatrix} 1-\lambda & 2 \\ 3 & 2-\lambda \end{bmatrix} = 0 \Rightarrow (1-\lambda)\cdot(2-\lambda) - (3)\cdot(2) = 0 \quad \boxed{\lambda_1 = -1 \; ; \; \lambda_2 = 4}$$

- Subs. back into eqns. or det. expression

$$\begin{bmatrix} 1 & 2 \\ 3 & 2 \end{bmatrix}\begin{bmatrix} u_x \\ u_y \end{bmatrix} = (-1)\begin{bmatrix} u_x \\ u_y \end{bmatrix} \Rightarrow \begin{cases} u_x + 2u_y = -u_x \\ 3u_x + 2u_y = -u_y \end{cases} \qquad \begin{bmatrix} 1 & 2 \\ 3 & 2 \end{bmatrix}\begin{bmatrix} u_x \\ u_y \end{bmatrix} = (4)\begin{bmatrix} u_x \\ u_y \end{bmatrix} \Rightarrow \begin{cases} u_x + 2u_y = 4u_x \\ 3u_x + 2u_y = 4u_y \end{cases}$$

choose $u_x = 1 \Rightarrow u_y = -1$ $\qquad\qquad$ choose $u_x = 1 \Rightarrow u_y = 3/2$

$$\boxed{\lambda_1 = -1 \; : \; u_1 = \begin{bmatrix} 1 \\ -1 \end{bmatrix}} \qquad \boxed{\lambda_2 = 4 \; : \; u_2 = \begin{bmatrix} 1 \\ 3/2 \end{bmatrix}}$$

The computation of eigenvalues and their corresponding eigenvectors for an N x N matrix **A** can be a daunting task requiring all the roots of an Nth degree polynomial. The solution for a simple 2 x 2 matrix is easily performed and illustrates the general solution procedure which may be then extended to higher dimensional cases. The eigenvalue-eigenvector equation states that the effect of **A** on an eigenvector **u** is to simply multiply **u** by a "scaling factor" λ yielding a parallel vector. Noting that multiplication of any vector by a unit matrix **I**=diag(1,1) leaves the vector unchanged allows us to write the second equality in the 1st equation at the top of the slide.

Re-writing this expression as a homogeneous matrix equation (**A**-λ **I**) **u** =0 leads us to conclude that a solution exist if and only if the determinant det(**A**-λ **I**) is zero. This in turn leads to a 2nd degree polynomial equation in λ whose roots are the desired eigenvalues. Substituting these eigenvalues back into the matrix equation one at a time allows us to solve for the two components of the eigenvector **u** corresponding to each eigenvalue. Note that since the determinant is zero, the two equations are not independent so we actually only have one equation for the two unknown components of **u**.

The eigenvalue polynomial is obtained by simply subtracting λ from each diagonal element and taking the determinant; the example 2 x 2 matrix yields two roots λ_1 = -1 and λ_2 = +4. Thus substituting λ_1 = -1 back into the top equation yields only a ratio u_y / u_x so assigning u_x = 1 we find u_y = -1 and a unit vector may be formed by dividing by $(1^2+(-1)^2)^{1/2}$ = $(2)^{1/2}$. Even though we have found both components, the solution is not unique as we could as well have taken u_y = 1 to find u_x = -1. The second eigenvector is found by substituting λ_2 = +4 and again assuming u_x = 1 we find u_y = 3/2. The complete solution is summarized in the boxed equations on the bottom of the slide.

7.3.4 Eigenvector-Eigenvalue Discussion

Eigenvector-Eigenvalue Discussion

- u_1, u_2 form a "eigen-basis"
  (*not necess. orthogonal or unit vectors*)
- Matrix **A** scales the components of **x** along u_1 & u_2 by $\lambda_1 = -1$ & $\lambda_2 = 4$ respectively.
- Eigen - Vector/Value may be complex
- Symmetric case $A = A^T$
  - Eigenvalues are real
  - Eigenvectors are orthogonal
- Non-orthogonal Decomposition of A

$$A = U\Lambda U^{-1} = \begin{bmatrix} 1 & 1 \\ -1 & 3/2 \end{bmatrix} \cdot \begin{bmatrix} -1 & 0 \\ 0 & 4 \end{bmatrix} \cdot \frac{2}{5}\begin{bmatrix} 3/2 & -1 \\ 1 & 1 \end{bmatrix}$$

$$= \frac{2}{5}\begin{bmatrix} 1 & 1 \\ -1 & 3/2 \end{bmatrix}\begin{bmatrix} -3/2 & 1 \\ 4 & 4 \end{bmatrix} = \frac{2}{5}\begin{bmatrix} 5/2 & 5 \\ 15/2 & 5 \end{bmatrix} = \begin{bmatrix} 1 & 2 \\ 3 & 2 \end{bmatrix}$$

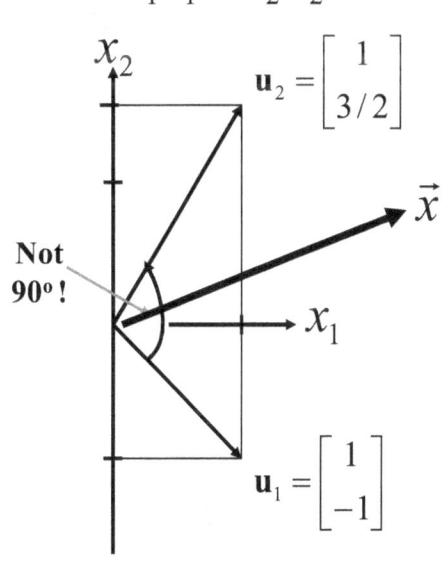

$$\mathbf{x} = x_1 \mathbf{u}_1 + x_2 \mathbf{u}_2$$

$$\mathbf{u}_2 = \begin{bmatrix} 1 \\ 3/2 \end{bmatrix}$$

Not 90°!

$$\mathbf{u}_1 = \begin{bmatrix} 1 \\ -1 \end{bmatrix}$$

It is easily verified by taking the scalar product that $\mathbf{u}_1^T \mathbf{u}_2 = 1 - 3/2 = -1/2 \neq 0$ and so the pair of eigenvectors in the example of the previous slide are in fact not orthogonal. This is shown by actually drawing them to scale in the figure on this slide. We note that the matrix **A** in that example was not symmetric having off diagonal elements of $a_{21} = 3$ and $a_{12} = 2$. This is no accident as it turns out that symmetry of **A** guarantees that the eigenvalues are real and the eigenvectors are orthogonal; thus for a symmetric matrix **A** we have an orthogonal spectral decomposition which gives a clean separation between A's effects upon on orthogonal components.

In general for a non-symmetric matrix, **A**, the eigenvalues are complex and the eigenvectors are non-orthogonal and **A** must be expressed in terms of **U** and its inverse $\mathbf{U}^{-1}$ (rather than the orthogonal spectral decomposition for a symmetric matrix given by the boxed equation on Slide#7-8.) The calculation to verify this explicitly is shown at the bottom of the slide.

7.3.5 Spectral Radius of a Matrix

Spectral Radius of a Matrix

- **Spectral Radius is max eigenvalue:** $\rho(A) = \max_{\text{all } k} |\lambda_k|$
- **Theorem: Spectral radius/Norm relationship**
  - (i) $\|\mathbf{A}\|_2 = \sqrt{\rho(\mathbf{A}^T\mathbf{A})}$
  - (ii) $\rho(\mathbf{A}) \leq \|\mathbf{A}\|$ for any natural norm
  - Corollary: $\rho(\mathbf{A})$ is a greatest lower bound (glb)

- **Example:**
  $$\mathbf{A} = \begin{bmatrix} 1 & 2 \\ 3 & 2 \end{bmatrix} \quad \mathbf{A}^T\mathbf{A} = \begin{bmatrix} 1 & 3 \\ 2 & 2 \end{bmatrix}\begin{bmatrix} 1 & 2 \\ 3 & 2 \end{bmatrix} = \begin{bmatrix} 10 & 8 \\ 8 & 8 \end{bmatrix} \rightarrow \det\begin{bmatrix} 10-\lambda & 8 \\ 8 & 8-\lambda \end{bmatrix} = 0 \; ; \; \lambda = 17.06, \, 0.94$$

  - Spectral radius of $\mathbf{A}^T\mathbf{A}$: $\rho(\mathbf{A}^T\mathbf{A}) = \max\{17.06, \, 0.94\} = 17.06$
  - 2-norm from Thm : $\|\mathbf{A}\|_2 = \sqrt{\rho(\mathbf{A}^T\mathbf{A})} = \sqrt{17.06} = 4.13$
  - Corollary: ρ is glb : $\rho(\mathbf{A}) = \max\{|-1|, |4|\} = 4$

  Direct Computation
  row/col sums of $\mathbf{A}$
  $$\begin{cases} \|\mathbf{A}\|_\infty = \max_{\text{row sums}} \begin{bmatrix} 1 & 2 \\ 3 & 2 \end{bmatrix} = \max_{\text{row sums}} \{3, 5\} = 5 \\ \|\mathbf{A}\|_1 = \max_{\text{col sums}} \{4, \, 4\} = 4 \end{cases}$$

  $\|\mathbf{A}\|_\infty = 5$
  $\|\mathbf{A}\|_2 = 4.13$
  $\|\mathbf{A}\|_1 = 4$

  glb: $\rho(\mathbf{A}) = 4$

The spectral radius of a matrix $\rho(\mathbf{A})$ is defined to be the eigenvalue with the maximum magnitude. The Theorem relating the spectral radius to the Euclidean 2-norm states that the 2-norm $\|\mathbf{A}\|_2$ is equal to the "square root" of the spectral radius of the matrix $\mathbf{A}^T\mathbf{A}$ (and not just $\mathbf{A}$), viz., $\|\mathbf{A}\|_2 = (\rho(\mathbf{A}^T\mathbf{A}))^{\frac{1}{2}}$ and that further the spectral radius $\rho(\mathbf{A})$ serves as a "greatest lower bound" for any of the natural matrix norms $\|\mathbf{A}\|_1$, $\|\mathbf{A}\|_2$, $\|\mathbf{A}\|_\infty$, etc.

The example illuminates the meaning of this theorem by computing the three norms and showing that they all are greater than or equal to the spectral radius of the matrix itself $\rho(\mathbf{A})$ as illustrated in the inset figure at the bottom right of the slide. The details of the computation are shown on the slide; the spectral radius of $\mathbf{A}$ by taking the max of the eigenvalues of the matrix $\mathbf{A}$: $\rho(\mathbf{A}) = \{|-1|, 4\} = 4$; the max-norm by taking the max of the absolute row sums of $\mathbf{A}$: $\|\mathbf{A}\|_\infty = \max\{1+2, 3+2\} = 5$; the 1-norm by taking the max of the absolute column sums of $\mathbf{A}$: $\|\mathbf{A}\|_1 = \max\{1+3, 2+2\} = 4$; finally, the Euclidean 2-norm by first computing $\mathbf{A}^T\mathbf{A}$, solving for its eigenvalues $\{17.06, 0.94\}$, obtaining its spectral radius and taking the square root $\|\mathbf{A}\|_2 = (\max\{17.06, 0.94\})^{\frac{1}{2}} = (17.06)^{\frac{1}{2}} = 4.13$. We see that all the norms fall into proper order and moreover, they all are $\geq \rho(\mathbf{A}) = 4$.

7.3.6 Convergent Matrix

Convergent Matrix

- **Defn:** Each element of kth power of matrix approaches zero $\mathrm{Lim}_{k \to \infty} (\mathbf{A}^k)_{ij} \to 0$ all i,j
- **Example:** $\mathbf{A}, \mathbf{A}^2, \mathbf{A}^3, \ldots \mathbf{A}^k, \ldots$

$$\mathbf{A} = \begin{bmatrix} 1/2 & 0 \\ 1/4 & 1/2 \end{bmatrix}; \quad \mathbf{A}^2 = \begin{bmatrix} 1/2 & 0 \\ 1/4 & 1/2 \end{bmatrix} \cdot \begin{bmatrix} 1/2 & 0 \\ 1/4 & 1/2 \end{bmatrix} = \begin{bmatrix} 1/4 & 0 \\ 1/4 & 1/4 \end{bmatrix}$$

$$\mathbf{A}^3 = \begin{bmatrix} 1/2 & 0 \\ 1/4 & 1/2 \end{bmatrix} \cdot \begin{bmatrix} 1/4 & 0 \\ 1/4 & 1/4 \end{bmatrix} = \begin{bmatrix} 1/8 & 0 \\ 3/16 & 1/8 \end{bmatrix}; \cdots, \mathbf{A}^k = \begin{bmatrix} (1/2)^k & 0 \\ k \cdot (1/2)^{k+1} & (1/2)^k \end{bmatrix} \Rightarrow \lim_{k \to \infty} (\mathbf{A}^k)_{ij} = 0$$

- **Theorem:**   **For Example above:**

Equivalent Statements:

(i) $\mathbf{A}$ is convergent

(ii) $\lim\limits_{k \to \infty} \|\mathbf{A}^k\| = 0$ one of the natural norms

(iii) $\lim\limits_{k \to \infty} \|\mathbf{A}^k\| = 0$ all natural norms

(iv) $\rho(\mathbf{A}) < 1$

(v) $\lim\limits_{k \to \infty} \|\mathbf{A}^k \mathbf{x}\| = 0$ all $\mathbf{x}$

$\det(\mathbf{A} - \lambda\mathbf{I}) = (1/2 - \lambda) \cdot (1/2 - \lambda) = 0 \to \lambda = 1/2, 1/2$

$\rho(\mathbf{A}) = \max\{1/2, 1/2\} = 1/2 < 1$

A matrix is said to be convergent if for all products of the matrix with itself $\{\mathbf{A}, \mathbf{AA}, \mathbf{AAA}, \ldots, \mathbf{A}^k\}$, every (i,j) element converges in the limit that the power $k \to \infty$; that is, $\lim_{k \to \infty} (\mathbf{A}^k)_{ij} \to 0$ for all elements (i,j). Multiplying the example matrix $\mathbf{A}$ by itself several times makes the pattern for each element clear and we see that each element has a factor of $(½)^k$ or $k(½)^{k+1}$ both of which approach 0 as $k \to \infty$. Thus the matrix $\mathbf{A}$ is convergent and further we calculate its eigenvalues to be $\{½, ½\}$, so that its spectral radius $\rho(\mathbf{A}) = \max\{½, ½\} = ½$.

The Theorem on this slide states the equivalence between the four statements (i) $\mathbf{A}$ is convergent, (ii) each element of the matrix to k^{th} power is convergent for **one of the natural norms**, (iii) if convergent for one of the natural norms (as in (ii)) then it converges for **all natural norms**, (iv) the spectral radius is less than "1", and (v) the matrix to k^{th} power matrix times an arbitrary $\mathbf{x}$ yields a new vector $\mathbf{y}$ which also converges.

Thus, for convergent matrices, we only need prove convergence of $\mathbf{A}$ using just one of the natural norms and we are assured that convergence holds for all natural norms. The fact that a convergent matrix has a spectral radius $\rho(\mathbf{A}) < 1$ is a result we shall need when we discuss theorems on the convergence of solutions to linear matrix equation $\mathbf{Ax} = \mathbf{b}$ (see Slide# 7-23). In the sequel, we shall see that the iterative solution of the matrix equation $\mathbf{Ax}=\mathbf{b}$ can be framed as $\mathbf{x}^{(k)} = \mathbf{T}\,\mathbf{x}^{(k-1)} + \mathbf{c}$, where the vector $\mathbf{c}$ is a constant and $\mathbf{T}$ depends upon $\mathbf{A}$. Thus, for $k \to k+1$, we have the next iterate

$$\mathbf{x}^{(k+1)} = \mathbf{T}\,\mathbf{x}^{(k)} + \mathbf{c} = \mathbf{T}\{\mathbf{T}\,\mathbf{x}^{(k-1)} + \mathbf{c}\} + \mathbf{c} = \mathbf{T}^2 \mathbf{x}^{(k-1)} + \mathbf{Tc} + \mathbf{c}$$

Thus we generate powers of a matrix $\mathbf{T}$ which must therefore be a convergent matrix.

7.4 Motivation for Iterative Methods

Motivation for Iterative Methods

- Direct Gauss Elimination $\sim 2n^3/3$ ops
    - Large system n=10,000 $\rightarrow$ 7 x 10^{11} ops
    - RO Error accumulates over 7 x 10^{11} ops
      & makes result meaningless
- Solutions to systems of diff eqns:
    - **A** is a sparse matrix: n~1000-100,000
    - Banded: use Special Direct Methods to reduce ops
    - Otherwise: need iterative methods
    - Even if need 1000 iterations fewer ops than direct
    - RO error in performing last iteration is all that counts!

The motivation for iterative methods is to control round off error for large system of equations. For a system with n=10,000 equations, the number of MAD operations needed for direct Gauss Elimination is $2/3\ n^3 \sim 7$ x 10^{11} and since the solution is not complete until the last step is performed, the round off error accumulates over this huge number of arithmetic operations which makes the result virtually useless.

Now, there are more efficient methods that apply to tri-diagonal, banded, sparse, circulant, and many other classes of matrices, which have repetitive, symmetric, sparse (few non-zero elements), or other special structures. Methods adapted to these specific structures allow significant reduction in the number of MAD operations by orders of magnitude. For example, in the case of a tri-diagonal matrix a special algorithm, which takes advantage of its simple banded structure, reduces the number of operations significantly to 8n-7 ~ 80,000 and will therefore produce accurate results.

For matrices without special structure the Gaussian Elimination procedure is replaced by an iterative procedure that starts with an initial guess for all n components of the solution vector $\mathbf{x}^{(0)}$ and produces a sequence of iterates $\{\mathbf{x}^{(0)}, \mathbf{x}^{(1)}, ..., \mathbf{x}^{(k)}\}$ which converge to the true solution **x**. Even if a large number of iterations is needed for convergence, say k=1000, the total number of operations for all the iterations $1000n^2$ can still be substantially less than the $2/3\ n^3$ required for Gaussian Elimination.

Unlike Gaussian Elimination, which accumulates errors over all its $2/3\ n^3$ operations before obtaining a solution vector, the iterative technique creates a new estimate of the solution vector at the end of each iteration, thus accumulating error only over $1*n^2$ operations. If the iterates are converging, then computing iterate $\mathbf{x}^{(k+1)}$ from iterate $\mathbf{x}^{(k)}$ is like starting over again, but with a "better initial guess". Thus, the round off error for a converging iterate starts "anew" in each iteration and, since all elements of the solution vector are obtained at the end of each iteration, only the round off that results from the last iteration counts!

Iterative Solution of Linear Systems

7.4.1 Formulation of Iterative Methods

Formulation of Iterative Methods

Functional Iteration (Fixed Point)

- Write N-explicit equations
  & Solve for diagonal element

$$a_{i1}x_1 + a_{i2}x_2 + \cdots + a_{ik}x_k + \cdots a_{iN}x_N = b_i$$

$$x_i = \underbrace{\frac{1}{a_{ii}}\left(b_i - \sum_{\alpha \neq i} a_{i\alpha}x_\alpha\right)}_{\equiv g_i(\bar{x})} \; ; \; i = 1,2,\cdots,N$$

Solve One component at a time

- Initial guess on RHS: $\quad x^{(0)} = [0,0,0,\cdots,0]^T$

- Yields iterates for each component on LHS

$$x_i^{(1)} = g_i(\bar{x}^{(0)}) \; ; \; i = 1,2,\cdots,N$$
$$x_i^{(2)} = g_i(\bar{x}^{(1)}) \; ; \; i = 1,2,\cdots,N$$
$$\vdots$$
$$x_i^{(k)} = g_i(\bar{x}^{(k-1)}) \; ; \; i = 1,2,\cdots,N$$

- Iterative Method ~ N^2 ops (not ~ N^3)

The iteration techniques for solution vectors $\mathbf{x}^{(k)}$ mirror those used for (scalar) root finding except that the norms of vector differences must now be used for convergence criteria of the vector sequence $\{\mathbf{x}^{(0)}, \mathbf{x}^{(1)}, ..., \mathbf{x}^{(k)}\}$. Functional iteration is the simplest technique and is accomplished by writing out the n equations implied by the matrix equation $\mathbf{A}\mathbf{x} = \mathbf{b}$ explicitly solving for the diagonal element in each row as given by the top equation for the solution component x_i, i=1,2, ...,N. Of course, any iteration procedure must start with an initial guess for all components of the solution vector, $\mathbf{x}^{(0)}$ which can be taken as "1"s or "0"s, or any vector that appears "natural".

Letting the function "g" denote the sum and difference operations of the top equation, we can write the functional iterations as shown in the 2nd set of equations. Thus the 1st iteration is performed on each indexed diagonal component i=1,2, ...,N and is written symbolically as $x_i^{(1)} = g(\mathbf{x}^{(0)})$ for the ith component. Evaluating g on the "vector" $\mathbf{x}^{(0)}$ symbolizes the specific operations of the top equation and updates the 1st iterate one-component at a time. The parenthesized superscript index represents the particular iterate and the lower index represents the component of that iterate; thus the symbol $x_i^{(k)}$ in the last equation represents the ith vector component of the kth iterate in the sequence.

There are a number of ways that this functional iteration can be accomplished and the next few slides give specific examples for a simple system of 2 equations which is most easily visualized.

Iterative Solution of Linear Systems

7.4.2 Jacobi – Simultaneous Displacement

Jacobi – Simultaneous Displacement

- **2x2 Linear System Example**

$$\left. \begin{array}{l} E1: 2x_1 + 1x_2 = +2 \\ E2: 1x_1 - 2x_2 = -2 \end{array} \right\} \rightarrow \begin{array}{l} x_1 = (2 - x_2)/2 \\ x_2 = (2 + x_1)/2 \end{array} \rightarrow \left. \begin{array}{l} x_1^{(k)} = (2 - x_2^{(k-1)})/2 \\ x_2^{(k)} = (2 + x_1^{(k-1)})/2 \end{array} \right\} \quad k = 1, 2, \cdots$$

$$\mathbf{x}^{(0)} = \begin{bmatrix} 0 \\ 0 \end{bmatrix} \qquad \mathbf{x}^{exact} = \begin{bmatrix} 0.4 \\ 1.2 \end{bmatrix}$$

$k = 1: \begin{cases} x_1^{(1)} = (2 - \underbrace{x_2^{(0)}}_{=0})/2 = 1 \\ x_2^{(1)} = (2 + \underbrace{x_1^{(0)}}_{=0})/2 = 1 \end{cases} \qquad \mathbf{x}^{(1)} = \begin{bmatrix} 1 \\ 1 \end{bmatrix}$

$k = 2: \begin{cases} x_1^{(2)} = (2 - \underbrace{x_2^{(1)}}_{=1})/2 = 1/2 \\ x_2^{(2)} = (2 + \underbrace{x_1^{(1)}}_{=1})/2 = 3/2 \end{cases} \qquad \mathbf{x}^{(2)} = \begin{bmatrix} 1/2 \\ 3/2 \end{bmatrix}$

$k = 3: \begin{cases} x_1^{(3)} = (2 - \underbrace{x_2^{(2)}}_{=3/2})/2 = 1/4 \\ x_2^{(3)} = (2 + \underbrace{x_1^{(2)}}_{=1/2})/2 = 5/4 \end{cases} \qquad \mathbf{x}^{(3)} = \begin{bmatrix} 1/4 \\ 5/4 \end{bmatrix}$

$k = 4: \begin{cases} x_1^{(4)} = (2 - \underbrace{x_2^{(3)}}_{=5/4})/2 = 3/8 \\ x_2^{(4)} = (2 + \underbrace{x_1^{(3)}}_{=1/4})/2 = 9/8 \end{cases} \qquad \mathbf{x}^{(4)} = \begin{bmatrix} 3/8 \\ 9/8 \end{bmatrix}$

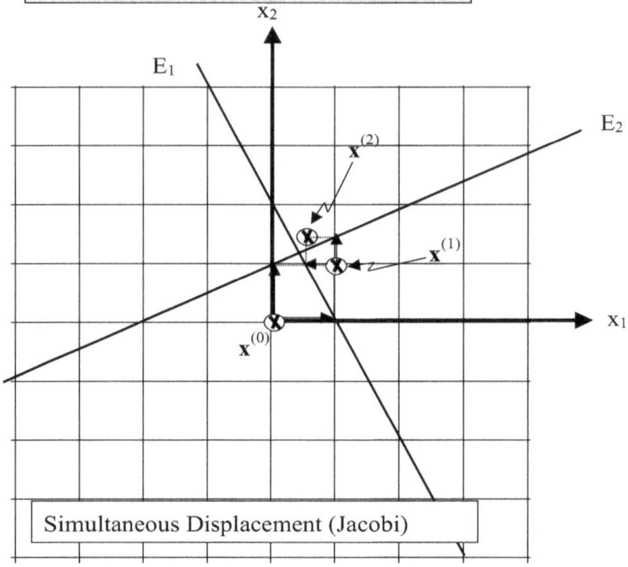

Simultaneous Displacement (Jacobi)

The system of two equations E1 and E2 are solved for their diagonal elements x_1 for E1 and x_2 for E2 and the boxed equation shows the two iteration equations obtained by placing superscript (k) representing the "updated iterate value" on the left hand side of the equation and the superscript (k-1) representing the "previous iterate value" on the right hand side. Starting with an initial guess $\mathbf{x}^{(0)} = [0,0]^T$ (at the origin), the iteration sequence is generated using the "update" equations and gives the sequence of vectors shown on the left.

The two equation lines labeled E1 and E2 in the figure intersect at the solution vector $\mathbf{x} = [0.4, 1.2]^T$; as we plot each iterate $\mathbf{x}^{(0)} = [0,0]^T$, $\mathbf{x}^{(1)} = [1,1]^T$, $\mathbf{x}^{(2)} = [1/2, 3/2]^T$, $\mathbf{x}^{(3)} = [1/4, 5/4]^T$, $\mathbf{x}^{(4)} = [3/8, 9/8]^T$ computed on the left, it is evident that the sequence in in fact converging (slowly) towards the solution vector $\mathbf{x} = [0.4, 1.2]^T$.

This is the Jacobi Method which starts at the origin $\mathbf{x}^{(0)} = [0,0]^T$ and makes "**simultaneous displacements**" $\Delta \mathbf{x}^{(1)} = \mathbf{x}^{(1)} - \mathbf{x}^{(0)} = [1, 1]^T$ along the x_1 and x_2 directions as indicated by the two dark arrows emanating from the origin and ends up at the 1st iterate vector position $\mathbf{x}^{(1)} = [1,1]^T$ as illustrated in the figure. The next iteration starts at $\mathbf{x}^{(1)} = [1,1]^T$ and makes the simultaneous displacements $\Delta \mathbf{x}^{(2)} = \mathbf{x}^{(2)} - \mathbf{x}^{(1)} = [-1/2, 1/2]^T$ indicated by the two black arrow emanating from the position $\mathbf{x}^{(1)} = [1,1]^T$ to arrive at the 2nd iterate vector position $\mathbf{x}^{(2)} = [1/2, 3/2]^T$. As this process continues the iterates are clearly "zeroing in on the true solution" and the sequence $\{\mathbf{x}^{(0)}, \mathbf{x}^{(1)}, ..., \mathbf{x}^{(k)}\}$ is seen to converge.

Iterative Solution of Linear Systems

7.4.3 Gauss-Seidel–Successive Displacement

Gauss-Seidel–Successive Displacement
- **2x2 Linear System Example**

$$\left.\begin{array}{l}E1: 2x_1 + 1x_2 = +2 \\ E2: 1x_1 - 2x_2 = -2\end{array}\right\} \rightarrow \begin{array}{l}x_1 = (2-x_2)/2 \\ x_2 = (2+x_1)/2\end{array} \rightarrow \begin{array}{l}x_1^{(k)} = (2-x_2^{(k-1)})/2 \\ x_2^{(k)} = (2+x_1^{(k)})/2\end{array}$$

"(k)" not "(k-1)"

$$\mathbf{x}^{(0)} = \begin{bmatrix}0\\0\end{bmatrix} \qquad \mathbf{x}^{exact} = \begin{bmatrix}0.4\\1.2\end{bmatrix}$$

$$x_1^{(1)} = (2 - \underbrace{x_2^{(0)}}_{=0})/2 = 1$$
$$x_2^{(1)} = (2 + \underbrace{x_1^{(1)}}_{=1})/2 = 3/2 \qquad \mathbf{x}^{(1)} = \begin{bmatrix}1\\3/2\end{bmatrix}$$

$$x_1^{(2)} = (2 - \underbrace{x_2^{(1)}}_{=3/2})/2 = 1/4$$
$$x_2^{(2)} = (2 + \underbrace{x_1^{(2)}}_{=1/4})/2 = 9/8 \qquad \mathbf{x}^{(2)} = \begin{bmatrix}1/4\\9/8\end{bmatrix}$$

$$x_1^{(3)} = (2 - \underbrace{x_2^{(2)}}_{=9/8})/2 = 7/16$$
$$x_2^{(3)} = (2 + \underbrace{x_1^{(3)}}_{=7/16})/2 = 39/32 \qquad \mathbf{x}^{(3)} = \begin{bmatrix}7/16\\39/32\end{bmatrix}$$

Faster Convergence !!

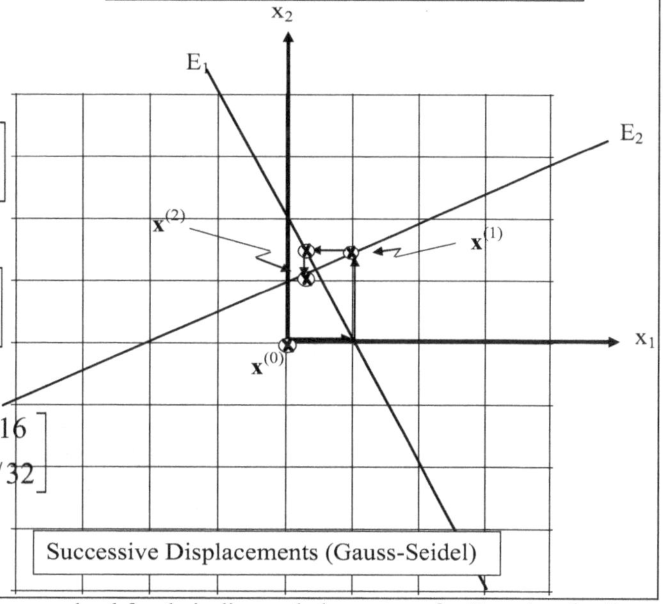

Successive Displacements (Gauss-Seidel)

Again the system of two equations E_1 and E_2 are solved for their diagonal elements x_1 for E_1 and x_2 for E_2 and the boxed equation shows the two iteration equations obtained by placing the superscript (k) representing the "updated iterate value" on the left hand side of the equation, but now instead of always using previous values, we use the most recent **updated values** for each component on the right hand side. Specifically, in the update equation for the 1st component $x_1^{(k)}$ we do not have any new updates yet so we use the "previous iterate value" superscript (k-1) on the right hand side as before; however when updating the 2nd component $x_2^{(k)}$ we make use of updated value we have just computed $x_1^{(k)}$ (not the old value $x_1^{(k-1)}$) on the right hand side as shown in the 2nd update equation.

Starting with an initial guess $\mathbf{x}^{(0)} = [0,0]^T$ (at the origin), the iteration sequence is generated using these new "update" equations which make use of the "most recently updated value for each component" and gives the sequence of vectors shown on the left. Note the (red) arrows showing, for example, that the updated value $x_1^{(1)}$ is used on the right hand side when computing the 2nd updated component $x_2^{(1)}$, etc. .

The two equation lines labeled E_1 and E_2 in the figure intersect at the solution vector $\mathbf{x}=[0.4, 1.2]^T$; as we plot each iterate $\mathbf{x}^{(0)} = [0,0]^T$, $\mathbf{x}^{(1)} = [1,3/2]^T$, $\mathbf{x}^{(2)} = [1/4, 9/8]^T$, $\mathbf{x}^{(3)} = [7/16, 39/32]^T$ computed on the left, it is evident that the sequence in in fact converging (actually faster than Jacobi) towards the solution vector $\mathbf{x}=[0.4, 1.2]^T$. This is the **Gauss-Seidel Method** which starts at the origin $\mathbf{x}^{(0)} = [0,0]^T$ and makes "**successive displacements**" $\Delta \mathbf{x}_1^{(1)} = \mathbf{x}_1^{(1)} - \mathbf{x}_1^{(0)} = [1, 0]^T$ along x_1 and then $\Delta \mathbf{x}_2^{(1)} = \mathbf{x}_2^{(1)} - \mathbf{x}_2^{(0)} = [0, 3/2]^T$ along x_2 directions as indicated by the two sequential dark arrows first from the origin 1-unit along x_1 and then from that point 3/2-unit along x_2 to arrive at the 1st iterate vector position $\mathbf{x}^{(1)} = [1,3/2]^T$ as illustrated in the figure. The next iteration starts at $\mathbf{x}^{(1)} = [1,1]^T$ and makes a similar set of successive displacements first along x_1 and then x_2 to arrive at the 2nd iterate vector position $\mathbf{x}^{(2)} = [1/4, 9/8]^T$. As this process continues, the iterates are clearly seen to "spiral in" and converge to the true solution. A very simple MatLab® script that just compares the converged values and number of iterates for the Jacobi and Gauss-Seidel methods is given on Slide#8-14.

7.4.4 Gauss-Seidel Divergence

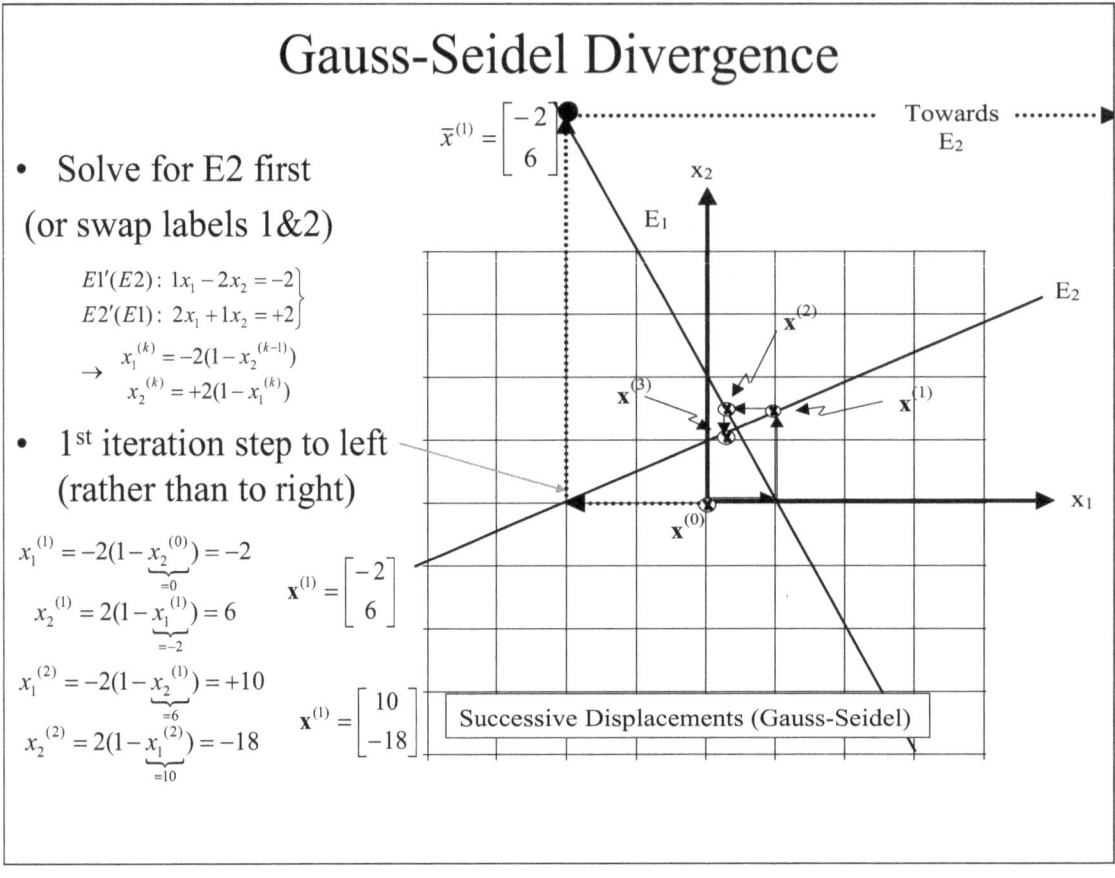

In Gaussian Elimination the order in which the equations are solved was found to affect the round off error; for iterative methods, solving the equations in the wrong order is also important as it can lead to divergence of the iterates.

Consider reversing the order of solution in our example by swapping the equations and relabeling them as E1'(old E2) and E2' (old E1) so that when solving for the first component of an iterate we are actually using the equation corresponding to E2. (Geometrically this means that the first displacement is towards the line E2 line rather than the E1 line.)

Starting from the same initial guess $x^{(0)} = [0,0]^T$, the iteration sequence is generated using these new (swapped) "update" equations as shown on the left yielding $x^{(1)} = [-2,6]^T$, $x^{(2)} = [10, -18]^T$ which is obviously diverging from the true solution vector $x=[0.4, 1.2]^T$. To understand what is happening we look at the figure to see how the sequence progresses from a geometric viewpoint.

The 1st component of iterate #1 is a displacement to the left (dashed arrow) 2-units to meet the line E1'(old E2) rather than to the right to meet the old E1 (solid arrow); the 2nd component then displaces the solution up 6-units to meet the line E2'(old E1) and yields the 1st iterate at location $x^{(1)} = [-2,6]^T$. Continuing in this manner, the 1st component of iterate #2 displaces the solution 10-units to the right to meet E1'(old E2) as indicated somewhere off the page; the 2nd component of iterate #2 displaces the solution down 18-units to the position of the 2nd iterate $x^{(2)} = [10, -18]^T$. As this process continues the iterates are clearly seen to "spiral outward" and diverge away from the true solution.

7.4.5 Successive Orthogonal Projections

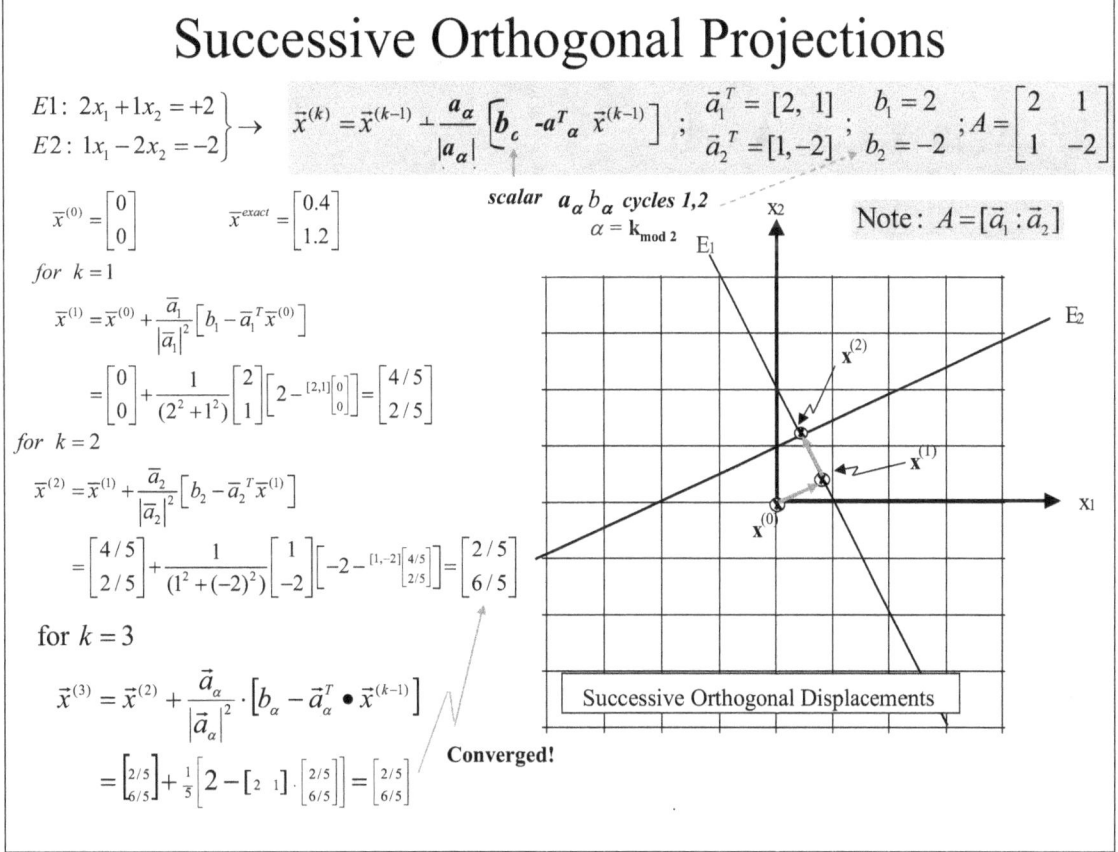

The two equations E1 and E2 are iterated as follows: starting with an initial guess $\mathbf{x}^{(0)}$ first drop a perpendicular to (the line determined by equation) E1 by setting k=1 in the iterate formula to yield the 1st iterate $\mathbf{x}^{(1)}$; next drop a perpendicular to E2 by setting k=2 in the iterate formula to yield the 2nd iterate $\mathbf{x}^{(2)}$; continue alternating between the two equations in this manner until convergence.

Note that in this iterate formula, all components of the kth iterate vector $\mathbf{x}^{(k)}$ are obtained using the vectorized iteration update equation; also the vector $\mathbf{a}_k$ occurs both as a column vector outside the square brackets and as a row vector $\mathbf{a}_k^T$ of the matrix $\mathbf{A}$ within the brackets where it forms a scalar product with the last iterate $\mathbf{x}^{(k-1)}$ column vector.

Starting with an initial guess $\mathbf{x}^{(0)} = [0,0]^T$ (at the origin), the iteration sequence is generated using the "update" equations and gives the sequence of vectors shown on the left. The two equation lines labeled E1 and E2 in the figure intersect at the solution vector $\mathbf{x}=[0.4, 1.2]^T$; as we plot each iterate $\mathbf{x}^{(0)} = [0,0]^T$, $\mathbf{x}^{(1)} = [4/5, 2/5]^T$, $\mathbf{x}^{(2)} = [2/5, 6/5]^T$ and the sequence converges to the exact solution after only two iterations. Note that this exact convergence after two iterations occurs because the two lines E1 and E2 are in fact perpendicular to each other; so traveling along the perpendicular to E1 from the origin to the line E1 makes the next iteration travel along the perpendicular to E2 which must end on E2 at their intersection.

The iteration formula holds for any number of dimensions and each iterate goes along one dimension at a time until all n iterates have been computed; the sequence is then repeated if needed. The perpendiculars to the two lines in the simple 2-dimensional case are replaced by normal to the planes defined by equation $\mathbf{Ax} = \mathbf{b}$; the kth hyper-plane is defined by $\mathbf{a}_k^T \mathbf{x}^{(k)} = b_k$, where $\mathbf{a}_k^T$ is the kth row vector of the row-partitioned n x n dimensional matrix $\mathbf{A}$. In order that the increment is only in the "hyper-plane" orthogonal to the direction $\mathbf{a}_k$ we require the projection operator $\mathbf{P}_k = [\mathbf{I} - \mathbf{a}_k \mathbf{a}_k^T / \|\mathbf{a}_k\|^2]$ acting on $\Delta\mathbf{x}^{(k)} = \mathbf{x}^{(k)} - \mathbf{x}^{(k-1)}$ to be zero, i.e., $\mathbf{P}_k \Delta\mathbf{x}^{(k)} = 0$. Rearranging terms, this leads to the equation $\mathbf{x}^{(k)} = [\mathbf{I} - \mathbf{a}_k \mathbf{a}_k^T / \|\mathbf{a}_k\|^2] \mathbf{x}^{(k-1)} + [\mathbf{a}_k (\mathbf{a}_k^T \mathbf{x}^{(k)}) / \|\mathbf{a}_k\|^2]$, and upon recognizing that $\mathbf{a}_k^T \mathbf{x}^{(k)} = b_k$ this yields the equation shown on the slide.

7.4.6 Diagonal Dominance & Convergence

Diagonal Dominance & Convergence

2^d Matrix Eqn Ax=b:

$$\left.\begin{array}{l}a_{11}x_1 + a_{12}x_2 = b_1 \\ a_{21}x_1 + a_{22}x_2 = b_2\end{array}\right\} \rightarrow \left.\begin{array}{l}x_1^{(k)} = (b_1 - a_{12}x_2^{(k-1)})/a_{11} \\ x_2^{(k)} = (b_2 - a_{21}x_1^{(k)})/a_{22}\end{array}\right\}$$

$$\Delta x_1^{(k)} = (\underbrace{\Delta b_1}_{=0} - a_{12}\Delta x_2^{(k-1)})/a_{11} = -a_{12}\Delta x_2^{(k-1)}/a_{11} \quad (1)$$

$$\Delta x_2^{(k)} = (\Delta b_2 - a_{21}\Delta x_1^{(k)})/a_{22} = -a_{21}\Delta x_1^{(k)}/a_{22} \quad (2)$$

$$k \rightarrow k-1: \quad \Delta x_2^{(k-1)} = -a_{21}\Delta x_1^{(k-1)}/a_{22}$$

$$\Delta x_1^{(k)} = \left(\frac{-a_{12}}{a_{11}} \cdot \frac{-a_{21}}{a_{22}}\right) \cdot \Delta x_1^{(k-1)} = \left(\frac{a_{12}}{a_{11}} \cdot \frac{a_{21}}{a_{22}}\right) \cdot \frac{-a_{12}}{a_{11}} \Delta x_2^{(k-2)} = \left(\frac{a_{12}}{a_{11}} \cdot \frac{a_{21}}{a_{22}}\right)^2 \Delta x_1^{(k-2)} = \cdots$$

$$= \left(\frac{a_{12}}{a_{11}} \cdot \frac{a_{21}}{a_{22}}\right)^k \Delta x_1^{(0)} \xrightarrow[k \rightarrow \infty]{} 0 \quad provided \quad \left|\left(\frac{a_{12}}{a_{11}} \cdot \frac{a_{21}}{a_{22}}\right)\right| < 1$$

$$\Rightarrow |a_{11}| > |a_{12}| \quad and \quad |a_{22}| > |a_{21}| \quad \text{Diagonal dominance}$$

Note this is a sufficient condition for convergence

We pointed out that, although Gauss-Seidel method converges more rapidly than the Jacobi method, care must be taken to solve the equations in the "correct order" to avoid potential divergence of the Gauss-Seidel iterates. However, the "correct order" needs to be defined so we can know ahead of time when to expect the iterates to converge; the key turns out to be related to the values of the diagonal terms in the system of equations and leads once again to the concept of diagonal dominance.

Consider the case n=2 for a system of just two equations and formally solve them successively for the diagonal terms in each row using the Gauss-Seidel "instant update scheme" as shown in the top equations for the two components of the solution vector $x_1^{(k)}$ and $x_2^{(k)}$. It we define the difference between two 1st component iterates by $\Delta x_1^{(k)} = x_1^{(k)} - x_1^{(k-1)}$ and similarly for the 2nd component iterates $\Delta x_2^{(k)} = x_2^{(k)} - x_2^{(k-1)}$, then we can take the '$\Delta$'s of the top two equations (or equivalently write them down twice with incremented indexes and subtract) to obtain the set numbered (1) and (2). These equations may be combined repeatedly by first using the second equation to give us $\Delta x_2^{(k-1)}$ in terms of $\Delta x_1^{(k-1)}$ for substitution back into Eq.(1) to obtain the first equality equating $\Delta x_1^{(k)}$ to a multiple $[(a_{12}/a_{11})(a_{21}/a_{22})] * \Delta x_1^{(k-1)}$ of the previous iterate. Repeating this process k times as indicated finally yields $\Delta x_1^{(k)}$ as a multiple to the kth power $[(a_{12}/a_{11})(a_{21}/a_{22})]^k * \Delta x_1^{(0)}$ of the difference between the first iterate and the initial guess. An analogous formula is obtained for the 2nd component $\Delta x_2^{(k)}$. Clearly if the iteration sequence is to converge the square bracket factor which is raised to the kth power must be less than unity. We note that it is sufficient for the separate ratios of the off-diagonal to diagonal terms (a_{12}/a_{11}) and (a_{21}/a_{22}) to *each be less than unity*, that is, for the matrix to be diagonally dominant. (It is of course possible that the product of the two is less than unity even if the individual ratios are not; but making each ratio less than unity certainly works and is thus "sufficient"). The extension of this proof to higher dimensions is more involved, but again leads to diagonal dominance as the sufficient condition for convergence.

7.4.7 Gauss-Seidel Convergence Theorems

Gauss-Seidel Convergence Theorems

- In each row i, $\quad a_{i1}x_1 + a_{i1}x_1 + \cdots + \widehat{a_{im}}x_m + \cdots a_{iN}x_N = b_i$
  solve for x_m that is has **dominant element** "a_{im}" which satisfies
- **Strictly Dominant matrix**: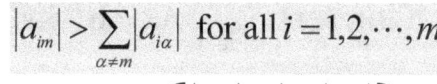
  has the dominant elements ■
  occuring in *different columns*
  for each row as shown by black ■
- Pivoting leads to diagonal dominance ▫
- **Theorem: Strict Dominance**
  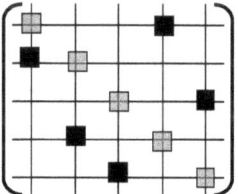
  - **If** Strictly Dominant & solve for dominant variable in each row
  - **Then** Gauss-Seidel(G-S) iterates converge
    - For arbitrary initial vector x_0
    - Indep of ordering of eqns.
- **Theorem: Positive Definite A**
  - If A is positive definite ($x^T A x > 0$ all nonzero x)
  - Then G-S iterates converge whether or not A is diagonally dominant

A strictly dominant matrix is one whose largest element in each row occurs in a different column in each row as indicated by the black squares in the matrix sketch. By a series of row swaps this strictly dominant form may be changed into a diagonally dominant form shown by the red boxes along the diagonal.

Here we state two theorems for the Gauss-Seidel method:

The first states that if the matrix is strictly dominant, then the Gauss-Seidel method converges for any initial vector provided we solve for the dominant element in each row, *i.e.*, there is no need to swap rows.

The second theorem states that if **A** is positive definite then Gauss-Seidel converges for any start vector and diagonal dominance is irrelevant, *i.e.*, positive definiteness "trumps" diagonal dominance.

7.5 Additive Matrix Decomposition

Additive Matrix Decomposition

Formal Matrix Decomposition

Convert: $\mathbf{Ax} = \mathbf{b} \rightarrow \mathbf{x}^{(k)} = \mathbf{Tx}^{(k-1)} + \mathbf{c}$; $k = 1, 2, \cdots$

Additive Decomposition: $\mathbf{A} = -\mathbf{L}_0 + \mathbf{D} - \mathbf{U}_0$

Where:

$$\mathbf{L}_0 = -\begin{bmatrix} 0 & 0 & 0 & 0 \\ l_{21} & 0 & 0 & 0 \\ l_{N-1,1} & \ddots & 0 & 0 \\ l_{N1} & l_{N2} & l_{N,N-1} & 0 \end{bmatrix} \quad \mathbf{D} = \begin{bmatrix} a_{11} & 0 & 0 & 0 \\ 0 & a_{22} & 0 & 0 \\ 0 & 0 & \ddots & 0 \\ 0 & 0 & 0 & a_{NN} \end{bmatrix} \quad \mathbf{U}_0 = -\begin{bmatrix} 0 & u_{12} & \cdots & u_{1N} \\ 0 & 0 & \cdots & u_{2N} \\ 0 & 0 & \ddots & u_{N-1,N} \\ 0 & 0 & 0 & 0 \end{bmatrix}$$

Substitute Decomposition for A $\quad (-\mathbf{L}_0 + \mathbf{D} - \mathbf{U}_0)\mathbf{x} = \mathbf{b}$

Formally write: $\quad \mathbf{Dx}^{(k)} = \mathbf{b} + \mathbf{L}_0 \mathbf{x}^{(k-1)} + \mathbf{U}_0 \mathbf{x}^{(k-1)}$

Solve for diagonal component $\quad \mathbf{x}^{(k)} = \underbrace{\mathbf{D}^{-1}(\mathbf{L}_0 + \mathbf{U}_0)}_{\equiv \mathbf{T}} \mathbf{x}^{(k-1)} + \underbrace{\mathbf{D}^{-1}\mathbf{b}}_{\equiv \mathbf{C}}$

Desired form: $\quad \mathbf{x}^{(k)} = \mathbf{Tx}^{(k-1)} + \mathbf{c}$

We have previously discussed how beneficial it is to factor the **A** matrix into products of special triangular and diagonal matrices; here we look at a additive decomposition and use it to characterize several different formulations of iterative methods within a common framework. The basic matrix equation **Ax =b** is converted to the common iterative form $\mathbf{x}^{(k)} = \mathbf{T}\,\mathbf{x}^{(k-1)} + \mathbf{C}$, where the matrix **T** is constructed from the original Matrix **A** and **C** is a constant vector constructed from **A** and **b**.

To show this we break up **A** into a lower triangular matrix **L₀** and an upper triangular matrix **U₀** both with zeros on their diagonals -**L₀** and –**U₀** and a diagonal matrix **D** as shown explicitly on the slide Thus we replace **Ax =b** with (-**L₀** + **D** –**U₀**)**x** = **b** and the iteration is formally written as shown on the slide; left multiplying by $\mathbf{D}^{-1}$ yields the Jacobi Method (simultaneous update) in the desired form $\mathbf{x}^{(k)} = \mathbf{T_J}\,\mathbf{x}^{(k-1)} + \mathbf{C_J}$, where we readily identify the form for $\mathbf{T_J} = \mathbf{D}^{-1}(\mathbf{L_0} + \mathbf{U_0})$ and $\mathbf{C_J} = \mathbf{D}^{-1}\mathbf{b}$. It is interesting to note that the Orthogonal Projection method is already in this form (see slide# 7-18).

7.5.1 Jacobi, Gauss-Seidel, Orthogonal Projection

Jacobi, Gauss-Seidel, Orthogonal Projection

- **Formally write:** $(-L_0 + D - U_0)x = b$

  - **Jacobi:** $Dx^{(k)} = b + L_0 x^{(k-1)} + U_0 x^{(k-1)}$

    $$x^{(k)} = \underbrace{D^{-1}(L_0 + U_0)}_{\equiv T_J} x^{(k-1)} + \underbrace{D^{-1}b}_{\equiv C_J}$$

    $$\boxed{x^{(k)} = T_J x^{(k-1)} + c_J}$$

  - **Gauss-Seidel:**

    k^{th} Iteration; i^{th} Step:

    $$\left\{ x_1^{(k)}\ x_2^{(k)}\ x_{i-1}^{(k)} : x_{i+1}^{(k-1)}\ x_{i+2}^{(k-1)} \cdots x_N^{(k-1)} \right\}$$

    use updated values ; only old values available

    $$a_{ii} x_i^{(k)} = b_i - \underbrace{\sum_{\alpha=1}^{i-1} a_{i\alpha} x_\alpha}_{i>\alpha} - \underbrace{\sum_{\alpha=i+1}^{N} a_{i\alpha} x_\alpha}_{i<\alpha}$$

    Step# i → Updated Values "k" (rows 1..i-1); Old "k-1" (rows i+1..N)

    $Dx^{(k)} = b + L_0 x^{(k)} + U_0 x^{(k-1)}$

    $(D - L_0) x^{(k)} = U_0 x^{(k-1)} + b$

    $$x^{(k)} = \underbrace{(D-L_0)^{-1} U_0}_{\equiv T_{GS}} x^{(k-1)} + \underbrace{(D-L_0)^{-1} b}_{\equiv C_{GS}}$$

    $$\boxed{x^{(k)} = T_{GS} x^{(k-1)} + c_{GS}}$$

  - **Orthogonal Projection**

    Note: $x^{(k)}$ & a_k are col. vectors

    $$\vec{x}^{(k)} = \underbrace{\left[I - \frac{\vec{a}_k \vec{a}_k^T}{|\vec{a}_k|^2} \right]}_{\equiv T_{OP}} \vec{x}^{(k-1)} + \underbrace{\frac{b_k}{|\vec{a}_k|^2} \vec{a}_k}_{\equiv C_{OP}}$$

    $$\boxed{x^{(k)} = T_{OP} x^{(k-1)} + c_{OP}}$$

    $\vec{a}_k = row_k(A) = [a_{k1}, a_{k2}, a_{k3}, \cdots a_{kN}]^T$

The *Jacobi* form is repeated here for convenience. In order to write down the Gauss-Seidel method in a convenient matrix form, we must do a step-by-step component update for each iterate making sure to use the most recent updates. In step #i of the kth iterate all components of the solution vector with index < i have already been updated and thus have values with superscript (k): $x_i^{(k)}$ as illustrated on the slide.

For the *Gauss-Seidel* form (middle panel) we write down the equations for the step#i of the kth iterate explicitly in terms of components and we see that the terms on the left with $a_{ii} x_i^{(k)}$ can be equated to the vector form $Dx^{(k)}$, while those over the first "already updated" partition with index $\alpha < i$ can be associated with the vector form $L_0 x^{(k)}$ and finally those with index $\alpha > i$ can be associated with the vector form $U_0 x^{(k-1)}$. Now collecting terms with the same iteration superscript $x_i^{(k)}$, the left hand side becomes $(D - L_0)x^{(k)}$ and subsequent left multiplication by the inverse $(D - L_0)^{-1}$ yields the boxed standard form in the middle panel boxed equation, where the matrix and constant for the *Gauss-Seidel* iterations are identified as $T_{GS} = (D - L_0)^{-1} U_0$ and $C_{GS} = (D - L_0)^{-1} b$ respectively.

Finally in the bottom panel, the *Orthogonal Projection* algorithm is easily put into the desired form by collecting together the two terms with $x^{(k-1)}$ in the boxed equation of Slide#7-18.

We shall see that there are several important theorems which compare the different methods when they are written in this general form.

7.5.2 Convergence and Spectral Radius $\rho(T)$

Convergence and Spectral Radius $\rho(T)$

- **Theorem: Convergence & $\rho(T)$**
  - **If and only if** the spectral radius of **T** satisfies: $\rho(T) < 1$
  - **Then** $x^{(k)} = Tx^{(k-1)} + c$ converges to unique solution for arbitrary $x^{(0)}$

- **Corol. on Error Bounds** (analogous to fixed point)

  If $\|T\| < 1$ for any natural norm, Then

  $(i)\ \|x - x^{(k)}\| \le \|T\|^k \cdot \|x - x^{(0)}\| \quad \xrightarrow{\rho(T) \approx \|T\|} \quad [\rho(T)]^k \cdot \|x - x^{(0)}\|$

  $(ii)\ \|x - x^{(k)}\| \le \dfrac{\|T\|^k}{1 - \|T\|} \cdot \|x^{(1)} - x^{(0)}\| \quad \xrightarrow{\rho(T) \approx \|T\|} \quad \dfrac{[\rho(T)]^k}{1 - \rho(T)} \cdot \|x^{(1)} - x^{(0)}\|$

  > Choose technique with smallest spectral radius $\rho(T)$

- **Theorem on Rate of Convergence (Stein-Rosenberg)**

  If (i) off $-$ diagonal neg. or zero: $a_{ij} \le 0$
  (ii) diagonal strictly positive: $a_{ii} > 0$
  Then one of following xor statements holds:

  (i) both converge: $0 \le \rho(T_{GS}) < \rho(T_J) < 1$ (GS faster)
  (ii) both diverge: $\rho(T_{GS}) > \rho(T_J) > 1$ (GS faster)
  (iii) both zero: $\rho(T_{GS}) = \rho(T_J) = 0$
  (iv) both unity: $\rho(T_{GS}) = \rho(T_J) = 1$

These theorems relate the convergence for all methods after casting them into the general iterative form $x^{(k)} = T\ x^{(k-1)} + C$, where the matrix **T** and the constant **C** are different for each formulation. The first theorem states that if the spectral radius $\rho(T)$ of a method is less than "1", then the method converges to a unique solution for any initial start vector $x^{(0)}$. (Note that on Slide#7-12 we found that a convergent matrix is equivalent to $\rho(T) < 1$ so we could just as well require **T** be a convergent matrix in the statement of the above theorem.)

The corollary to the above theorem gives error bounds that are analogous to those previously found for fixed point iterates in root finding. The only difference is that the bound on the derivative of the fixed point function $|g'(x)| < k < 1$ for the scalar iterates is now replaced by the spectral radius $\rho(T)$ for the vector iterates. Observe that the two error estimates actually involve the matrix norm $\|T\|$, but since the spectral radius is a lower bound for any norm (and is usually very close to all norms), the error estimates are re-cast in terms of $\rho(T)$.

We have seen that Gauss-Seidel generally converges faster than Jacobi, but there is no general result that this will always be the case. The second theorem compares the convergence rates of these two methods for a very restrictive set of conditions, namely that (i) off diagonal elements of **A** are negative semi-definite $a_{ij} \le 0$ and (ii) diagonal elements of **A** are strictly positive $a_{ii} > 0$. For an **A** matrix satisfying these restrictions, the spectral radius for the two methods $\rho(T_{GS})$ and $\rho(T_J)$ satisfy one and only one of the four mutually exclusive relations shown. In particular, the first two conditions mean that they both converge or diverge together with the GS method both always converging faster and always diverging faster.

7.6 Accelerating Convergence

Accelerating Convergence

1. Aitken's Method

– Track ratio of corrections to i^{th} component to satisfy criterion

$$\frac{\Delta_k \hat{x}_i^{(k)}}{\Delta_k \hat{x}_i^{(k-1)}} \cong const. \quad where, \quad \Delta_k = difference\ on\ iteration\ index\ "k"$$

– Accelerate by applying Aitken's "operator" to iteration 'k'

$$\hat{x}_i^{(k)}\Big)_{New} = \hat{x}_i^{(k)} - \frac{\left[\Delta_k \hat{x}_i^{(k)}\right]^2}{\Delta_k^2 \hat{x}_i^{(k)}} = \hat{x}_i^{(k)} - \frac{\left[\hat{x}_i^{(k)} - \hat{x}_i^{(k-1)}\right]^2}{\hat{x}_i^{(k)} - 2\hat{x}_i^{(k-1)} + \hat{x}_i^{(k-2)}} \quad ; i = 1, 2, \cdots, N$$

2. Successive Over Relaxation (SOR) Methods

– Write down following identity for the Gauss-Seidel iterates

$$\hat{x}_i^{(k)} \equiv \hat{x}_i^{(k-1)} + (\hat{x}_i^{(k)} - \hat{x}_i^{(k-1)}) = \hat{x}_i^{(k-1)} + \Delta \hat{x}_i^{(k)} \quad ; \quad \Delta \hat{x}_i^{(k)} \equiv (\hat{x}_i^{(k)} - \hat{x}_i^{(k-1)})$$

– Modify by making partial corrections for each iteration so as not to completely satisfy the "row" equations

$$\hat{x}_i^{(k)} = \hat{x}_i^{(k-1)} + \omega \cdot \Delta \hat{x}_i^{(k)} \quad ; \quad \begin{cases} \omega > 1 \quad over\ relaxation \\ \omega < 1 \quad under\ relaxation \end{cases}$$

$$\hat{x}^{(k)} = (D - \omega L_0)^{-1} [(1-\omega)D - \omega U_0] \hat{x}^{(k-1)} + \omega \cdot (D - \omega L_0)^{-1} \overline{b}$$

Two techniques that can be used to accelerate the convergence of vector iterates are (1) Aitken's Method applied to the scalar components, and (2) SOR relaxation methods which scale the vector corrections in an *ad hoc* manner.

Aitken's Method tracks the ratio of the changes in individual components (index "i") of successive iterates $\Delta x_i^{(k)} = x_i^{(k)} - x_i^{(k-1)}$ just as in root finding and when this ratio $\Delta x_i^{(k)} / \Delta x_i^{(k-1)}$ approaches a constant (*i.e.*, linear convergence), the algorithm switches to the Aitken's improved iterate for the ith component. The algorithm can track each element individually and switch or more efficiently track just a few of the largest elements and then switch to the Aitken's improved iterate simultaneously for all elements.

Successive Over-Relaxation (SOR) Method takes the identity

$$x_i^{(k)} = x_i^{(k)} + (x_i^{(k-1)} - x_i^{(k-1)}) = x_i^{(k-1)} + (x_i^{(k)} - x_i^{(k-1)}) = x_i^{(k-1)} + \Delta x_i^{(k)}$$

in which the kth iterate $x_i^{(k)}$ is written as the $(k-1)^{st}$ iterate $x_i^{(k-1)}$ plus an increment $\Delta x_i^{(k)}$ and basically considers not making the full increment, but rather a larger or smaller increment $\omega \Delta x_i^{(k)}$. Choosing a value $\omega > 1$ leads to larger increments and is called over relaxation, while $\omega < 1$ leads to smaller increments is called under relaxation. Generally, this choice is made in a somewhat *ad hoc* manner for a particular class of problems by testing a range of values of ω to determine which value requires the least number of iterations. This method can be written in the standard form $\mathbf{x}^{(k)} = T_\omega \mathbf{x}^{(k-1)} + \mathbf{C}_\omega$ as shown in the last equation on the slide and using this form it can be shown that ω must satisfy the inequality $0 < \omega < 2$ for convergence.

Iterative Solution of Linear Systems

7.6.1 SOR – Math Details

SOR – Math Details

Write down following identity for the Gauss-Seidel iterates
$$a_{ii}x_i^{(k)} = a_{ii}(x_i^{(k-1)} + \text{``1''} \cdot \Delta x_i^{(k)}) \quad ; \quad \Delta x_i^{(k)} \equiv (x_i^{(k)} - x_i^{(k-1)})$$

Replacing the "1" by "ω" destroys identity
$$a_{ii}x_i^{(k)} \cong a_{ii}x_i^{(k-1)} + \omega \cdot a_{ii}\Delta x_i^{(k)} = (1-\omega)a_{ii}x_i^{(k-1)} + \omega a_{ii}x_i^{(k)}$$

Substitute G-S value for $\mathbf{Dx}^{(k)}$ on RHS **only**
$$\mathbf{Dx}^{(k)} \cong (1-\omega)\mathbf{Dx}^{(k-1)} + \omega \cdot \mathbf{Dx}^{(k)} \quad \textit{Matrix-Vector Form}$$
$$\mathbf{Dx}^{(k)} = \mathbf{b} + \mathbf{L}_0 \mathbf{x}^{(k)} + \mathbf{U}_0 \mathbf{x}^{(k-1)} \quad \textit{Gauss-Seidel Method}$$

Yields
$$\mathbf{Dx}^{(k)} \cong (1-\omega)\mathbf{Dx}^{(k-1)} + \omega\left[\mathbf{b} + \mathbf{L}_0 \mathbf{x}^{(k)} + \mathbf{U}_0 \mathbf{x}^{(k-1)}\right]$$

Collecting terms and replacing $\mathbf{x}^{(k)}$ with $\hat{\mathbf{x}}^{(k)}$ restores the equality
$$(\mathbf{D} - \omega\mathbf{L}_0)\hat{\mathbf{x}}^{(k)} = ((1-\omega)\mathbf{D} + \omega\mathbf{U}_0)\hat{\mathbf{x}}^{(k-1)} + \omega\mathbf{b}$$

Finally, the SOR equations are given by pre-multiplying by the inverse $(\mathbf{D} - \omega\mathbf{L}_0)^{-1}$
$$\hat{\mathbf{x}}^{(k)} = \underbrace{(\mathbf{D} - \omega\mathbf{L}_0)^{-1}((1-\omega)\mathbf{D} + \omega\mathbf{U}_0)}_{=\mathbf{T}_\omega}\hat{\mathbf{x}}^{(k-1)} + \underbrace{\omega(\mathbf{D} - \omega\mathbf{L}_0)^{-1}\mathbf{b}}_{=\mathbf{C}_\omega}$$
$$\hat{\mathbf{x}}^{(k)} = \mathbf{T}_\omega \hat{\mathbf{x}}^{(k-1)} + \mathbf{C}_\omega$$

The Gauss-Seidel iterations work sequentially on each component (index i) and essentially zeros out the residual error in the dimension corresponding to the index "i" by traveling all the way to the surface with normal corresponding to Cartesian coordinate "x_i". The SOR method relaxes this zero residual constraint for component "i" by either falling short or traveling beyond the "x_i -coordinate surface."

Accordingly, the identity corresponding to the Gauss-Seidel iterate $x_i^{(k)}$ (shown in the top equation of the slide) is "relaxed" by replacing "1" $\Delta x_i^{(k)}$ with "ω" $\Delta x_i^{(k)}$. The definition $\Delta x_i^{(k)} = x_i^{(k)} - x_i^{(k-1)}$ is then re-inserted to yield the approximate equation on the 2nd line of the slide. The SOR formulation is most easily derived by casting the component equation into *matrix-vector form* as follows
$$\mathbf{Dx}^{(k)} \cong (1-\omega)\mathbf{Dx}^{(k-1)} + \omega \cdot \mathbf{Dx}^{(k)} \quad (1)$$

Next the Gauss-Seidel iterate given below
$$\mathbf{Dx}^{(k)} = \left[\mathbf{b} + \mathbf{L}_0 \mathbf{x}^{(k)} + \mathbf{U}_0 \mathbf{x}^{(k-1)}\right] \quad (2)$$
is substituted on the right hand side of Eq.(1) to yield
$$\mathbf{Dx}^{(k)} \cong (1-\omega)\mathbf{Dx}^{(k-1)} + \omega \cdot \left[\mathbf{b} + \mathbf{L}_0 \mathbf{x}^{(k)} + \mathbf{U}_0 \mathbf{x}^{(k-1)}\right] \quad (3)$$
Collecting all $\mathbf{x}^{(k)}$ terms of Eq.(3) on the left and all $\mathbf{x}^{(k-1)}$ terms on the right, and finally left multiplying (the resulting equation) by the matrix inverse $(\mathbf{D}-\omega\mathbf{L}_0)^{-1}$, we obtain the SOR equations in the standard form with $\mathbf{T}_\omega$ and $\mathbf{C}_\omega$ defined in the last boxed equation on the slide.

7.6.2 Example SOR Problem

Example SOR Problem

Form Gauss-Seidel Iterate from Original Equations

$$\left.\begin{array}{l} E1: 2x_1 + 1x_2 = +2 \\ E2: 1x_1 - 2x_2 = -2 \end{array}\right\} \rightarrow \begin{array}{l} x_1 = (2 - x_2)/2 \\ x_2 = (2 + x_1)/2 \end{array} \rightarrow \boxed{\begin{array}{l} x_1^{(k)} = 1 - x_2^{(k-1)}/2 \\ x_2^{(k)} = 1 + x_1^{(k)}/2 \end{array}}$$

Form SOR Eqns: Insert "ω" into identity $x_1^{(k)} = x_1^{(k-1)} + \underbrace{\Delta_{(k)} x_1^{(k)}}_{(?)}$ **and expand out again**

$$x_1^{(k)} = x_1^{(k-1)} + \omega \cdot \underbrace{\Delta_{(k)} x_1^{(k)}}_{x_1^{(k)} - x_1^{(k-1)}} = (1-\omega) \cdot x_1^{(k-1)} + \omega \cdot x_1^{(k)} \xrightarrow{\text{SOR}} x_1^{(k)} = (1-\omega)x_1^{(k-1)} + \omega \underbrace{\left(1 - \frac{1}{2} x_2^{(k-1)}\right)}_{=x_1^{(k)}}$$

$$x_2^{(k)} = x_2^{(k-1)} + \omega \cdot \underbrace{\Delta_{(k)} x_2^{(k)}}_{x_2^{(k)} - x_2^{(k-1)}} = (1-\omega) \cdot x_2^{(k-1)} + \omega \cdot x_2^{(k)} \xrightarrow{\text{SOR}} x_2^{(k)} = (1-\omega)x_2^{(k-1)} + \omega \underbrace{\left(1 + \frac{1}{2} x_1^{(k)}\right)}_{=x_2^{(k)}}$$

Note: $\omega = 1$ reduces to Gauss-Seidel

Here is the SOR set-up procedure with our example system of two equations; instead of using the boxed equation of the last slide it is simpler to actually follow the steps we took to "derive" the SOR matrix vector equation in the first place.

Specifically we proceed as if doing GS iterates shown in the top boxed pair of equations. Next we compute the "iterate delta" for component#1 by substituting update equation for $x_1^{(k)}$ into the definition $\Delta x_1^{(k)} = x_1^{(k)} - x_1^{(k-1)}$ to obtain an expression $\Delta x_1^{(k)} = (1-x_2^{(k-1)}/2) - x_1^{(k-1)}$. In a similar manner we find component#2 $\Delta x_2^{(k)} = (1+x_1^{(k)}/2) - x_2^{(k-1)}$; finally, substituting these values for $\Delta x_1^{(k)}$ and into $\Delta x_2^{(k)}$ into the "relaxed equations" in the bottom boxed equations, we obtain the explicit form of the SOR equations.

Note that following the "derivation steps" avoids the tortuous matrix inverses and multiplies of the formal expression on the last slide. Of course we have no idea what value to take for the relaxation parameter ω but it is bound in the interval (0,2) and we shall use $\omega = 1.2$ in the sample calculation on the next slide.

7.6.3 Explicit SOR Computation

Explicit SOR Computation

Iter#0 $\bar{x}^{exact} = \begin{bmatrix} 0.4 \\ 1.2 \end{bmatrix}$ $\bar{x}^{(0)} = \begin{bmatrix} 0 \\ 0 \end{bmatrix}$ Choose $\omega = 1.2$

Iter#1
$$x_1^{(1)} = \underbrace{(1-\omega) \cdot x_1^{(0)}}_{-0.2 \quad 0} + \underbrace{\omega \cdot (1 - x_2^{(0)}/2)}_{1.2 \quad =0} = 1.2$$
$$x_2^{(1)} = \underbrace{(1-\omega) \cdot x_2^{(0)}}_{-0.2 \quad 0} + \underbrace{\omega(1 + x_1^{(1)}/2)}_{1.2 \quad =1.2} = 1.92$$
$\bar{x}^{(1)} = \begin{bmatrix} 1.2 \\ 1.92 \end{bmatrix}$

Iter#2
$$x_1^{(2)} = (-0.2) \cdot \underbrace{x_1^{(1)}}_{1.2} + (1.2) \cdot (1 - \underbrace{x_2^{(1)}}_{1.92}/2) = -.192$$
$$x_2^{(2)} = (-0.2) \cdot \underbrace{x_2^{(1)}}_{1.92} + (1.2) \cdot (1 + \underbrace{x_1^{(2)}}_{-.192}/2) = .7008$$
$\bar{x}^{(2)} = \begin{bmatrix} -.192 \\ .7008 \end{bmatrix}$

Iter#3
$$x_1^{(3)} = (-0.2) \cdot \underbrace{x_1^{(2)}}_{-.192} + (1.2) \cdot (1 - \underbrace{x_2^{(2)}}_{.7008}/2) = .81792$$
$$x_2^{(3)} = (-0.2) \cdot \underbrace{x_2^{(2)}}_{.7008} + (1.2) \cdot (1 + \underbrace{x_1^{(3)}}_{.81792}/2) = 1.550592$$
$\bar{x}^{(3)} = \begin{bmatrix} .81792 \\ 1.550592 \end{bmatrix}$

Here we use the SOR equations in the form developed on the last slide and for a relaxation parameter $\omega = 1.2$. The steps are exactly the same as for Gauss-Seidel; it is just that the equations have been altered slightly by the introduction of a relaxation parameter ω. Note that this particular choice of relaxation parameter does not improve the convergence; in fact the 1st three iterations appear to be wandering rather than converging to the true solution.

Thus this algorithm should be used with caution because it may not work as well as the Jacobi or Gauss-Seidel methods for a particular set of equations. In this regard, see the following slides (Slide#7-28 and 7-29) which explore this issue of choosing a value for the SOR relaxation constant ω *via* theorems and simulations.

7.6.4 Choice of Relaxation Constant ω

Choice of Relaxation Constant ω
Theorems for Special Cases

– Thm(Kahan): ***Non-zero diag elem***	If $a_{ii} \neq 0$ for all i Then $\rho(T_\omega) \geq \|\omega - 1\|$ $\Rightarrow \boxed{\rho(T_\omega) < 1 \text{ only if } 0 < \omega < 2}$
– Thm(Ostrowski-Reich) ***Pos. def. A***	If A is pos. def. & $0 < \omega < 2$ Then SOR Method $\boxed{\text{converges for arb. initial } \bar{x}^{(0)}}$
– Thm(Tridiagonal) SOR / Jacobi Relationship ***Tridiagonal***	If A is pos. def. & Tridiagonal, Then (i) Spectral Radius $\boxed{\rho(T_{GS}) = [\rho(T_J)]^2 < 1}$ (ii) Optimal Choice for ω is $\boxed{\omega_{opt} = \dfrac{2}{1 + \sqrt{1 - \rho(T_{GS})}}}$ & $\rho(T_\omega) = \omega - 1$

As we found out in the last slide the SOR method comes with no guarantees about convergence nor any clue as to what values of the relaxation parameter to choose. Here are three theorems which give results for some specific types of matrices.

(1) The 1st theorem (Kahan) requires $a_{ii} \neq 0$, (**non-zero diagonal elements**) and puts the previously mentioned bounds $0 < \omega < 2$ on the relaxation parameter. The inequality $\rho(T) \geq |\omega - 1|$ does not guarantee $\rho(T) < 1$, but for there to be a chance that $\rho(T) < 1$ it is *absolutely necessary* that $|\omega - 1| < 1$; nonetheless convergence is still not guaranteed because this only ensures that the spectral radius is greater than a number that is *less than* 1. Thus, we require the absolute value inequality $|\omega - 1| < 1$ to be satisfied. The bounds on ω are found by separating this inequality into its two distinct parts,

(i) $\omega > 1$ which yields $\omega - 1 < 1$ or $\omega < 2$, and
(ii) $\omega < 1$ which yields $1 - \omega < 1$ or $\omega > 0$

(2) The 2nd theorem (Ostrowski-Reich) states a *sufficient condition* for the SOR method to converge with arbitrary start vector $x^{(0)}$ is that (i) the matrix **A** be positive definite and also (ii) that the relaxation parameter satisfies the bounds $0 < \omega < 2$. Note that our example matrix of the last slide is **not positive definite** so even though $\omega = 1.2$ is within its bounds, the iterates do not converge; moreover, this theorem only gives sufficient conditions for convergence. They are not necessary, as clearly for $\omega = 1.0$ we have Gauss-Seidel iterations which we have seen converge even though the matrix is not positive definite.

(3) The 3rd theorem on tridiagonal matrices **A** actually gives specific information on the optimal choice for the relaxation parameter as well as between the spectral radii of the Gauss-Seidel and Jacobi methods under the conditions that **A** is both tridiagonal and positive definite.

These theorems can be helpful in understanding what is going wrong and perhaps pointing towards a corrective procedure. However, in most circumstances they are not adequate both because they deal with special types of matrices and because they only state sufficient conditions for their conclusions which allows convergence even when their prerequisites are violated.

7.6.5 Number of Iterations *vs.* SOR Parameter ω

Number of Iterations *vs.* SOR Parameter ω

Linear System:
A is Pos. Def & Tridiagonal

$$A = \begin{bmatrix} 4 & 3 & 0 \\ 3 & 4 & -1 \\ 0 & -1 & 4 \end{bmatrix} \; ; \; b = \begin{bmatrix} 24 \\ 30 \\ -24 \end{bmatrix}$$

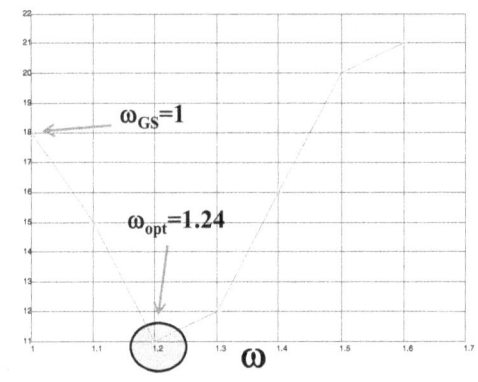

Decomposition of A: $A = D - L_0 - U_0$

$$D = \begin{bmatrix} 4 & 0 & 0 \\ 0 & 4 & 0 \\ 0 & 0 & 4 \end{bmatrix} \; ; \; L_0 = -\begin{bmatrix} 0 & 0 & 0 \\ 3 & 0 & 0 \\ 0 & -1 & 0 \end{bmatrix} \; ; \; U_0 = -\begin{bmatrix} 0 & 3 & 0 \\ 0 & 0 & -1 \\ 0 & 0 & 0 \end{bmatrix}$$

Iteration Equation: $\mathbf{x}^{(k)} = \mathbf{T}_\omega \mathbf{x}^{(k-1)} + \mathbf{c}_\omega$

$$\mathbf{T}_\omega = (D - \omega L_0)^{-1}[(1-\omega)D + \omega U_0]$$

$$\mathbf{c}_\omega = \omega \cdot (D - \omega L_0)^{-1} \mathbf{b}$$

Stopping Criterion
$\|\mathbf{p}_{i+1} - \mathbf{p}_i\|_2 < \varepsilon_{TOL} = 10^{-4}$

Optimal SOR parameter ω:

$$T_{GS} = (D - L_0)^{-1} U_0 = \begin{bmatrix} 4 & 0 & 0 \\ 3 & 4 & 0 \\ 0 & -1 & 4 \end{bmatrix}^{-1} \begin{bmatrix} 0 & -3 & 0 \\ 0 & 0 & 1 \\ 0 & 0 & 0 \end{bmatrix} = \frac{1}{64}\begin{bmatrix} 16 & 0 & 0 \\ -12 & 16 & 0 \\ -3 & 4 & 16 \end{bmatrix}\begin{bmatrix} 0 & -3 & 0 \\ 0 & 0 & 1 \\ 0 & 0 & 0 \end{bmatrix} = \begin{bmatrix} 0 & -48/64 & 0 \\ 0 & 36/64 & 16/64 \\ 0 & 9/64 & 4/64 \end{bmatrix}$$

Spectral Radius = Max Eigenvalue (T_{GS})

$$0 = \det(T_{GS} - \lambda I) = -\lambda\{(36/64 - \lambda)(4/64 - \lambda) - 9(16)/64\} \Rightarrow \lambda_{max} = \max\{0, 0, 40/64\} = 40/64$$

$$\omega_{opt} = \frac{2}{1 + \sqrt{1 - \rho(T_{GS})}} = \frac{2}{1 + \sqrt{1 - 40/64}} = 1.24$$

The theorem on the previous slide placed bounds on the relaxation constant as $0 < \omega < 2$. Here we study the effects of changing ω within these bounds for the 3-dimensional linear system $\mathbf{Ax} = \mathbf{b}$ given on the slide. The MatLab® script used to produce this plot along with a detailed listing of the iterate output tables for a subset of 5 values ω = [1.0, 1.1, 1.2, 1.3, 1.4] is given on Slide# 8-15. The 2-norm stopping criterion for these iterations is given by $\|\mathbf{p}_{i+1} - \mathbf{p}_i\|_2 < 10^{-4}$ and the plot clearly indicates an optimal value of n=10 iterations for ω near 1.2; away from this value of ω, there is a sharp rise in the number of iterations required for convergence.

Recall from the formulation of the SOR method that the value of ω is considered to be a scale factor that gauges the closeness to the Gauss-Seidel full correction to a given vector component at each step. Hence the case ω = 1.0 actually corresponds to a full correction, *i.e.*, the Gauss-Seidel method itself and we see from the plot that GS takes n=18 iterations for convergence which is nearly twice the n=10 iterations for convergence using the optimal value of ω =1.24.

The matrix **A** is *tridiagonal*, has positive diagonal elements and is diagonally dominant. These conditions are sufficient to to prove that A is *positive definite*, so the 3rd Theorem of the previous Slide#7-28 applies; the computation on the slide yields a spectral radius $\rho(T_{GS})$ = 40/64 and ω_{opt} = 1.24 in agreement with the simulation results.

Higher dimensional linear systems and a more stringent stopping criterion $\varepsilon_{TOL}=10^{-8}$ will tend to make this choice of optimal ω even more critical. As an example dependence on dimensions, the MatLab® script and tables given on Slides# 8-16, 8-17 compares Jacobi, Gauss-Seidel, SOR (for ω =1.2), and Orthogonal Projection Methods, for a *2-dimensional linear system*. In this case it turns out that the number of iterations for the same convergence criterion $\|\mathbf{p}_{i+1} - \mathbf{p}_i\|_2 < 10^{-4}$ are respectively n_J=15, n_{GS}=9, $n_{\omega=1.2}$ >20 (has not converged yet) and n_{OP}=2, so dimension truly is relevant to SOR convergence. A close look at the Table on Slide# 8-17 shows that both components of the SOR solution for ω =1.2 are trapped in a cycle oscillating between two values, but neither is converging with 5 significant digits.

Aside from the specific types of matrix **A** addressed by the theorems of the previous slide there is no way to determine the optimal value of ω other than to study its effect on a given class and dimension of linear system. This should be no surprise inasmuch as the introduction of ω was *ad hoc* in the first place!

Iterative Solution of Linear Systems

7.6.6 Comparison of Iterative Methods: Jacobi, Gauss-Seidel, SOR

Comparison of Iterative Methods: Jacobi, Gauss-Seidel, SOR

$$A = \begin{bmatrix} 4 & 3 & 0 \\ 3 & 4 & -1 \\ 0 & -1 & 4 \end{bmatrix} \; ; \; b = \begin{bmatrix} 24 \\ 30 \\ -24 \end{bmatrix} \qquad A = D - L_0 - U_0$$

$$D = \begin{bmatrix} 4 & 0 & 0 \\ 0 & 4 & 0 \\ 0 & 0 & 4 \end{bmatrix} \; ; \; L_0 = -\begin{bmatrix} 0 & 0 & 0 \\ 3 & 0 & 0 \\ 0 & -1 & 0 \end{bmatrix} \; ; \; U_0 = -\begin{bmatrix} 0 & 3 & 0 \\ 0 & 0 & -1 \\ 0 & 0 & 0 \end{bmatrix}$$

SOR $\omega = 1.25$

$$x^{(k)} = T_\omega x^{(k-1)} + c_\omega$$

$$T_\omega = (D - \omega L_0)^{-1}\left[(1-\omega)D + \omega U_0\right]$$

$$b_\omega = \omega \cdot (D - \omega L_0)^{-1} b$$

iter	Jacobi			Gauss-Seidel			SOR $\omega=1.25$		
0	1	1	1	1	1	1	1	1	1
1	5.25	7	-5.75	5.25	3.8125	-5.0469	6.3125	3.5195	-6.6501
2	0.75	2.125	-4.25	3.1406	3.8828	-5.0293	2.6223	3.9585	-4.6004
3	4.4063	5.875	-5.4688	3.0879	3.9268	-5.0183	3.1333	4.0103	-5.0967
4	1.5938	2.8281	-4.5313	3.0549	3.9542	-5.0114	2.9571	4.0075	-4.9735
5	3.8789	5.1719	-5.293	3.0343	3.9714	-5.0072	3.0037	4.0029	-5.0057
6	2.1211	3.2676	-4.707	3.0215	3.9821	-5.0045	2.9963	4.0009	-4.9983
7	3.5493	4.7324	-5.1831	3.0134	3.9888	-5.0028	3	4.0003	-5.0003
8	2.4507	3.5422	-4.8169	3.0084	3.993	-5.0017	2.9997	4.0001	-4.9999
9	3.3433	4.4578	-5.1144	3.0052	3.9956	-5.0011	3	4	-5
10	2.6567	3.7139	-4.8856	3.0033	3.9973	-5.0007	3	4	-5
11	3.2146	4.2861	-5.0715	3.002	3.9983	-5.0004			
12	2.7854	3.8212	-4.9285	3.0013	3.9989	-5.0003			
13	3.1341	4.1788	-5.0447	3.0008	3.9993	-5.0002			
14	2.8659	3.8882	-4.9553	3.0005	3.9996	-5.0001			
15	3.0838	4.1118	-5.0279	3.0003	3.9997	-5.0001			
16	2.9162	3.9302	-4.9721	3.0002	3.9998	-5			
17	3.0524	4.0698	-5.0175	3.0001	3.9999	-5			
46	2.9999	3.9999	-5						
47	3	4.0001	-5						
48	3	4	-5						
49	3	4	-5						
50	3	4	-5						

Jacobi

$$x^{(k)} = T_J x^{(k-1)} + c_J$$

$$T_J \equiv D^{-1}(L_0 + U_0)$$

$$c_J = D^{-1} b$$

Gauss-Seidel

$$x^{(k)} = T_{GS} x^{(k-1)} + c_{GS}$$

$$T_{GS} \equiv (D - L_0)^{-1} U_0$$

$$c_{GS} \equiv (D - L_0)^{-1} b$$

Stopping Criterion

$$\|p_{i+1} - p_i\|_2 < \varepsilon_{TOL} = 10^{-4}$$

Here is a comparison of the Jacobi, Gauss-Seidel and SOR ($\omega_{optimal} = 1.25$) methods for the same linear system $Ax = b$ used for the SOR study on the Slide#7-29. The MatLab® script on Slides#8-16, 8-17 with the alternate inputs (commented out) generates the table on this slide using the same 2-norm stopping criteron $\|p_{i+1} - p_i\|_2 < 10^{-4}$. The column layout has the iteration index i in col#1, and then the three vector components for Jacobi, Gauss-Seidel, and SOR across the table. SOR is the clear winner converging in 9 iterations, Gauss-Seidel in 16, and Jacobi in 48.

Note that we have used the general formalism developed in Slide#7-21 which makes use of the additive decomposition of the matrix $A = (-L_0 + D - U_0)$ to express all methods in the common iterative form $x^{(k)} = T x^{(k-1)} + c$. The matrix T and the constant vector c are constructed according to their definitions given in Slides# 7-22 and 7-25. The matrices L_0, D, and U_0 are written down explicitly in the slide and the definitions for T for each method are repeated there for convenience. This common formalism approach provides a very simple way to compare the methods and allows quick studies of how each approach works as we change the dimension of the linear system and the desired accuracy. We have seen that both factors affect the way these algorithms perform. .

7.6.7 Advantages of Iterative Algorithms

Advantages of Iterative Algorithms

- **Smaller Number of Operations per Iteration**
  - Direct Gauss Elimination $\sim 2n^3/3$ ops
  - Gauss-Seidel $\sim 2n^2$ ops

  Break-even: n/3 iterations

- **G-S has better RO characteristics**
  - Gauss Elim: has RO at each step
  - Gauss-Seidel: only *last iteration* RO applies

  G-S can withstand more ops during iteration before RO deteriorates to Gauss elimination level

- **G-S Algorithm Implementation**
  - Update stored value using current values at each step i
  - Storage Replacement

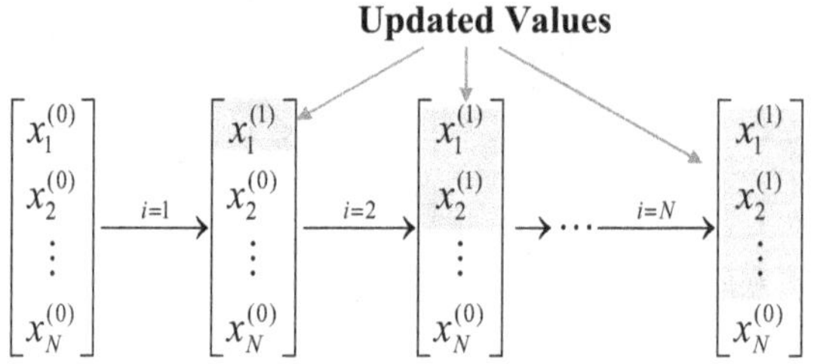

$$\begin{bmatrix} x_1^{(0)} \\ x_2^{(0)} \\ \vdots \\ x_N^{(0)} \end{bmatrix} \xrightarrow{i=1} \begin{bmatrix} x_1^{(1)} \\ x_2^{(0)} \\ \vdots \\ x_N^{(0)} \end{bmatrix} \xrightarrow{i=2} \begin{bmatrix} x_1^{(1)} \\ x_2^{(1)} \\ \vdots \\ x_N^{(0)} \end{bmatrix} \rightarrow \cdots \xrightarrow{i=N} \begin{bmatrix} x_1^{(1)} \\ x_2^{(1)} \\ \vdots \\ x_N^{(0)} \end{bmatrix}$$

(Updated Values)

$$\text{for } i = 1, \cdots, N \quad x_i \leftarrow \frac{1}{a_{ii}} \left(b_i - \sum_{\alpha \neq i} a_{i\alpha} x_\alpha \right)$$

A summary of the advantages of iterative over direct algorithms is given on this slide. First and foremost, the number of MAD operations for Gauss-Seidel is $\sim 2n^2$ compared with $\sim 2n^3/3$ for direct Gaussian Elimination; even if Gauss-Seidel took n/3 iterations yielding the same number of operations for each method, it still remains superior because the RO error only accumulates over the $2n^2$ operations required for the last iteration while Gauss accumulates RO error over all $2n^3/3$ operations.

It is also noteworthy that Gauss-Seidel is superior in terms of storage requirements because the updated value at each component stage of a given iterate can be updated in its appropriate vector storage location as shown in the illustration. This works because the update procedure always uses the most recent value and that is efficiently placed in the ith storage location always using the same n x 1 "storage" vector $\mathbf{x}^{(k)}$. The algorithm can be stated quite simply as follows "for i=1, ..., N, compute the update for x_i and place it in its unique storage location." The equation description on the bottom of the slide states the same thing mathematically - the arrow indicates the "assignment" of the updated value to the ith component of the storage vector.

7.7 Matrix Condition Number – Gaussian Elimination Issues

Matrix Condition Number– Gaussian Elimination Issues

- **Estimated solution** $\hat{x}$ to linear system $\quad Ax = b \quad (1)$
  - Subs. Back into original eqn. yields: $\hat{b} \equiv A\hat{x}$
  - So define data residual vector: $\Delta b \equiv b - \hat{b} = b - A\hat{x}$
  - And state error vector: $\Delta x \equiv x - \hat{x}$
- **Question:**
  - If data residuals Δb are small, are the state errors Δx small as well?
  - How does this depend upon the measurement matrix A?

System	"Bad" State Estimate	Data & State Residuals
$\begin{bmatrix} 1 & 2 \\ 1.0001 & 2 \end{bmatrix} \begin{bmatrix} x_1 \\ x_2 \end{bmatrix} = \begin{bmatrix} 3.0000 \\ 3.0001 \end{bmatrix}$ $x_{exact} = \begin{bmatrix} 1 \\ 1 \end{bmatrix}$	$\hat{x} = \begin{bmatrix} 3.0000 \\ .00000 \end{bmatrix} \Rightarrow$	$\Delta \bar{x} = \begin{bmatrix} 1 \\ 1 \end{bmatrix} - \begin{bmatrix} 3.0000 \\ .00000 \end{bmatrix} = \begin{bmatrix} -2.0000 \\ 1.0000 \end{bmatrix}$ $\Delta \bar{b} = \begin{bmatrix} 3.0000 \\ 3.0001 \end{bmatrix} - \begin{bmatrix} 1 & 2 \\ 1.0001 & 2 \end{bmatrix} \begin{bmatrix} 3.0000 \\ .00000 \end{bmatrix} = \begin{bmatrix} .00000 \\ -.00002 \end{bmatrix}$

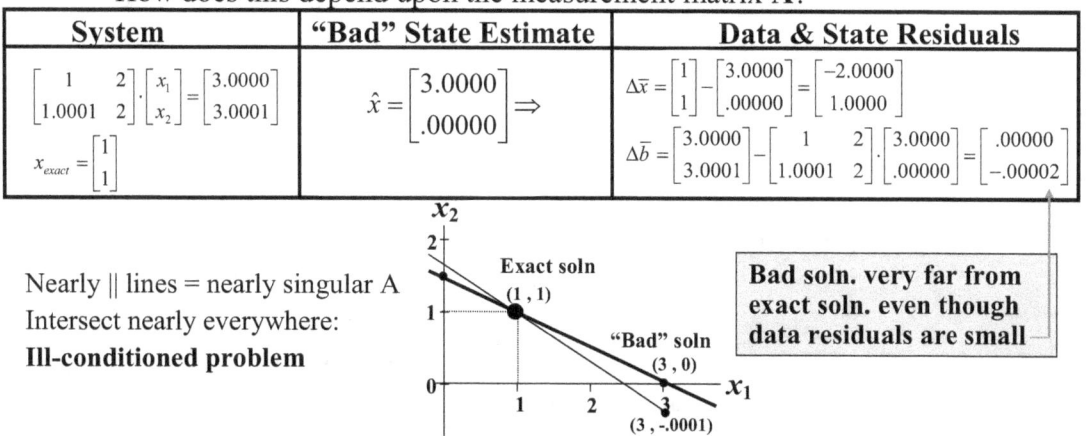

Nearly || lines = nearly singular A
Intersect nearly everywhere:
Ill-conditioned problem

Bad soln. very far from exact soln. even though data residuals are small

The equation $Ax = b$ can be thought of as one that produces the measured data b by a linear transformation on some fundamental state parameters x characterizing the object being observed. The transformation matrix in this context is the *measurement matrix* which relates the unknown state x to the observed data outputs b. Thus we are interested in the accuracy of our *state estimate* as measured by the vector difference $\Delta x = x - x_{hat}$ whose norm measures the estimation error.

After a Gaussian Elimination procedure we arrive at an estimate for the state vector which may have significant errors both because of the large number of operations and the nature of the matrix A itself. The action of A on this estimate gives a "predicted" value for the measurement $b_{hat} = A\, x_{hat}$ which is not exactly equal the true vector b. The difference vector $\Delta b = b - b_{hat}$ is a data residual vector and we would expect that small values for its norm is evidence that the state estimate x_{hat} is accurate. However, this is not always the case; as the simple example on this slide shows, the fact that the data residual norm $\|\Delta b\|$ is small does not necessarily mean that the norm of the state error $\|\Delta x\|$ is also small

The example system given in the 1st column of the table has an exact solution of $x_{exact} = [1,1]^T$ as is easily verified by direct multiplication of A times $[1,1]^T$ gives the data vector $b = [3.0000, 3.0001]^T$ exactly. Consider the following "bad" solution vector $x_{hat} = [3.0000, .00000]^T$ and look at the resulting data and state residuals computed in the 3rd column of the table. Even though the data residual vector $\Delta b = [.00000, -.00002]^T$, is in fact quite small, the state error vector is unacceptably large $\Delta x = [-2.0000, 1.0000]^T$.

Plotting the two equations shows that they intersect at a very shallow angle, and hence are "nearly parallel" and approximately *intersect everywhere*. The bad solution $[3.0000, .00000]^T$ is exactly on the first line E1: $1x_1 + 2 x_2 = 3.0000$, but is not on the second line E2: $1.0001x_1 + 2 x_2 = 3.0001$; for $x_1 = 3.0000$ the line E2 requires $x_2 = -.0001$ which lies slightly below the x_1-axis as shown. Comparing the slope and intercept of the two straight lines directly again shows that they are nearly on top of one another; this fact is easy to discern for a simple case, but in general, we need a method to determine such situations without resorting to graphical or analytic results.

7.7.1 Condition Number Definition

Condition Number Definition

- **Norms measure size of N-dimensional vectors**
  - Desire Ratio: norm of $\Delta \mathbf{x}$ to norm of $\Delta \mathbf{b}$
  - Data residuals satisfy: $\Delta \mathbf{b} = \underbrace{\mathbf{b}}_{=\mathbf{Ax}} - \mathbf{A}\hat{\mathbf{x}} = \mathbf{A}\Delta \mathbf{x}$ (2)
  - Solving (2) for $\Delta \mathbf{x} = \mathbf{A}^{-1}\Delta \mathbf{b}$ (2')
  - Taking norms of Eq.(1) [prev. slide] & Eq. (2') & multiplying them yields

  $$\left.\begin{array}{l} \|\mathbf{b}\| = \|\mathbf{Ax}\| \leq \|\mathbf{A}\|\cdot\|\mathbf{x}\| \quad (1) \\ \|\Delta\mathbf{x}\| = \|\mathbf{A}^{-1}\Delta\mathbf{b}\| \leq \|\mathbf{A}^{-1}\|\cdot\|\Delta\mathbf{b}\| \quad (2') \end{array}\right\} \Rightarrow \|\mathbf{b}\|\cdot\|\Delta\mathbf{x}\| \leq \{\|\mathbf{A}\|\cdot\|\mathbf{x}\|\}\cdot\{\|\mathbf{A}^{-1}\|\cdot\|\Delta\mathbf{b}\|\}$$

  - The norm ratio is defined by

  $$\frac{\|\Delta\mathbf{x}\|}{\|\mathbf{x}\|} \leq K(A)\cdot\frac{\|\Delta\mathbf{b}\|}{\|\mathbf{b}\|} \quad ; \quad K(A) \equiv \|\mathbf{A}\|\cdot\|\mathbf{A}^{-1}\|$$
  Condition Number

  $K(A)$ is *'force multiplier'* between %changes in data residuals and state errors

We have Eq.(1) $\mathbf{Ax} = \mathbf{b}$ for the true values and $\mathbf{Ax}_{hat} = \mathbf{b}_{hat}$ for the estimate, so subtracting these two equations yields Eq.(2) $\Delta\mathbf{b} = \mathbf{A}\,\Delta\mathbf{x}$ between the data residuals and the state errors. We seek a relation of the type shown in the boxed equation at the bottom of the slide which essentially tells us how a percentage change in the data residuals relates to a percentage change in the state error.

Taking the norms of Eqs.(1) and (2) we find $\|\mathbf{b}\| = \|\mathbf{Ax}\| \leq \|\mathbf{A}\|\,\|\mathbf{x}\|$ and $\|\Delta\mathbf{b}\| = \|\mathbf{A}\,\Delta\mathbf{x}\| \leq \|\mathbf{A}\|\,\|\Delta\mathbf{x}\|$. We might think to obtain a relation between the ratios $\|\Delta\mathbf{x}\| / \|\mathbf{x}\|$ and $\|\Delta\mathbf{b}\| / \|\mathbf{b}\|$ by dividing these two inequalities, but the results of that division are indeterminate. This is easily seen by the counter example: $3 \leq 4$ and $1 \leq 4$ would lead to the false statement $3/1 \leq 4/4$.

Thus, we need to take products of inequalities in the same order to obtain same order inequalities. Therefore invert Eq.(2) for $\Delta\mathbf{x} = \mathbf{A}^{-1}\Delta\mathbf{b}$, take its norm to obtain Eq.(2') $\|\Delta\mathbf{x}\| = \|\mathbf{A}^{-1}\|\,\|\Delta\mathbf{b}\|$ and finally take the product of Eq.(2') with the norm of Eq(1) to obtain the inequality shown on the slide. Solving for the ratio $\|\Delta\mathbf{x}\| / \|\mathbf{x}\|$ we find the boxed equation at the bottom, where the "condition number" $K(A) = \|\mathbf{A}\|\,\|\mathbf{A}^{-1}\|$ represents the multiplier between the two ratios.

Proof of the inequality $\|\mathbf{Ax}\| \leq \|\mathbf{A}\|\,\|\mathbf{x}\|$. This inequality is easily proved by first noting that any vector $\mathbf{x}$ can be written as its *norm* times a *unit vector*, viz., $\mathbf{x} = \|\mathbf{x}\|\cdot\mathbf{u}$, so we have

$$\|\mathbf{Ax}\| = \|\mathbf{A}(\|\mathbf{x}\|\mathbf{u})\| = \|\mathbf{x}\|\cdot\|\mathbf{Au}\|$$
$$\|\mathbf{Ax}\| \leq \|\mathbf{x}\|\cdot\max_{\|\mathbf{u}\|=1}\{\|\mathbf{Au}\|\}$$

Note that we have taken the scalar $\|\mathbf{x}\|$ outside of the norm expression on the first line and taken the maximum value of $\{\|\mathbf{Au}\|\}$ to give the inequality on the second line. Now, recalling the definition of the natural norm, viz., $\|\mathbf{A}\| = \max_{\|\mathbf{u}\|=1}\{\|\mathbf{Au}\|\}$, where $\mathbf{u}$ is an arbitrary *unit vector*, the inequality above becomes

$$\|\mathbf{Ax}\| \leq \|\mathbf{x}\|\,\|\mathbf{A}\|. \quad \text{QED}$$

Iterative Solution of Linear Systems

7.7.2 Estimates of Condition Number

Estimates of Condition Number

Example   **L₁\U**   **Gaussian Calc.**   **Scale**

$$\mathbf{A} = \begin{bmatrix} 1 & 2 \\ 1.0001 & 2 \end{bmatrix} \quad \mathbf{A}^{-1} = \begin{bmatrix} -10{,}000 & 10{,}000 \\ 5000.5 & -5000 \end{bmatrix} \quad [\mathbf{A}] \rightarrow [\mathbf{L}_1 \setminus \mathbf{U}] = \begin{bmatrix} (1) & 2 \\ 1.0001 & (-.0002) \end{bmatrix} \quad \bar{s} = \begin{bmatrix} 2 \\ 2 \end{bmatrix}$$

$\|A\|_\infty = \max\{3, 3.0001\} = 3.0001$; $\|A^{-1}\|_\infty = \max\{10{,}000, 20{,}000\} = 20{,}000$

Condition number estimates

1. $K(\mathbf{A}) = \|\mathbf{A}^{-1}\| \|\mathbf{A}\|$ any natural norm

   $K_\infty(\mathbf{A}) = \|\mathbf{A}\|_\infty \cdot \|\mathbf{A}^{-1}\|_\infty = 60{,}000$

2. $K_{\text{conte}}(\mathbf{A}) = \Pi(s_i) / \Pi(u_{ii})$
   Prod(Row Scale Factors)/ Prod(diag(L\U))

   $K_{\text{Conte}}(\mathbf{A}) = \dfrac{2 \cdot 2}{(1) \cdot |(-.0002)|} = 20{,}000$

3. $K_{\text{pivot}}(\mathbf{A}) = s_N / |u_{NN}|$
   (last row scale factor)/(last diag(L\U))

   $K_{\text{Pivot}}(\mathbf{A}) = \dfrac{2}{|(-.0002)|} = 10{,}000$

- **Rules-of-Thumb:**

  1. Computing with d significant digits the data residual error ratio is

     $\dfrac{\|\Delta \mathbf{b}\|}{\|\hat{\mathbf{b}}\|} \approx 10^{-d}$

  2. Using values of $\|\hat{\mathbf{x}}\|$ and $\|\Delta \mathbf{x}\| = \mathbf{A}^{-1} \|\Delta \mathbf{b}\|$ computed during Gaussian Elimination

     we estimate K(A) from: $\dfrac{\|\Delta \mathbf{x}\|}{\|\hat{\mathbf{x}}\|} \leq K(A) \cdot \dfrac{\|\Delta \mathbf{b}\|}{\|\hat{\mathbf{b}}\|} \leq K(A) \cdot 10^{-d} \Rightarrow K(A) \geq 10^{+d} \cdot \dfrac{\|\Delta \mathbf{x}\|}{\|\hat{\mathbf{x}}\|}$

The calculation of the condition number according to its definition is computationally expensive because it requires that we first find $\mathbf{A}^{-1}$, then compute product of the norms of $\mathbf{A}$ and $\mathbf{A}^{-1}$. Since the condition number is only used for diagnostic purposes or as a sentinel to warn of potential numerical issues, an approximate value for K(A) is usually sufficient.

Other approximations to the condition number are related to pivoting scale vectors and the diagonal elements generated in the Gaussian elimination procedure for solving the linear system $\mathbf{A}\mathbf{x} = \mathbf{b}$, as illustrated in the slide. One estimate $K_{\text{conte}} = P(s_i) / P(u_{ii})$ takes the ratio of products of row scale factors s_i to the product of diagonal elements of $\mathbf{U}$ in the $\mathbf{L}_1\backslash\mathbf{U}$ decomposition while the other simply takes the ratio $K_{\text{pivot}} = s_N / |u_{NN}|$ of the last row scale factor and the last diagonal element of $\mathbf{U}$. Calculations using each approximation are given for the simple n=2 system on the slide and we find $K_\infty(\mathbf{A}) = 60{,}000$, $K_{\text{conte}}(\mathbf{A}) = 20{,}000$, and $K_{\text{pivot}}(\mathbf{A}) = 10{,}000$.

Although the two approximations differ by factors of 3 and 6 from the $K_\infty(\mathbf{A})$ value, any of these values tells us that the condition number is quite large and that some special attention is necessary; the two approximations for the condition number are obtained using quantities that are already computed during the course of the Gaussian elimination and so they require *no additional intensive computational burden and should therefore always be calculated*.

Two rules of thumb are given at the bottom of the slide:
(1) The first states that when computing with d - digits of precision, the relative error in the data residuals is

$$\|\Delta \mathbf{b}\| / \|\hat{\mathbf{b}}\| = \|\mathbf{b} - \hat{\mathbf{b}}\| / \|\hat{\mathbf{b}}\| \approx 10^{-d} \quad \text{(loss of significant digits)}$$

Thus, if we have a condition number $K(\mathbf{A}) > 10^3$ and compute with 7 digits of precision, we lose 3 digits in the state error ratio accuracy, *i.e.*, the error occurs in the 7-3= 4th digit. (Note that all precision is lost for $K(\mathbf{A}) > 10^7$.)
(2) The second follows from substitution of the above data error ratio, (1), into the definition of K(A):
$(\|\Delta \mathbf{x}\|/\|\mathbf{x}\|) = K(A) \cdot (\|\Delta \mathbf{b}\|/\|\mathbf{b}\|)$ and to yield

$$K(\mathbf{A}) > 10^d \|\Delta \mathbf{x}\| / \|\hat{x}\|,$$

where d = # digits precision, the vector $\hat{x}$ is the approximate solution obtained by Gaussian Elimination, and $\Delta \mathbf{x}$ is obtained by solving $\mathbf{A} \Delta \mathbf{x} = \Delta \mathbf{b}$ **using the same $\mathbf{L}_1\backslash\mathbf{U}$ Gaussian factorization of A.** Again, most of the computation has been done in the factorization of $\mathbf{A}$ and thus these calculations allow good *sentinels* that are *almost free*.

7.8 Iterative Refinement for Gaussian Elimination

Iterative Refinement for Gaussian Elimination

- **Use Iterative Refinement when**
  - $K(\mathbf{A})$ is large ($\mathbf{A}$ is ill-conditioned)
  - N is large and RO error is likely to occur for large # ops
- **Procedure:**
  1. Gauss [L\U] w/pivoting
  2. Compute State Estimate Fwd/Bkwd Subs. $\hat{\mathbf{x}}$
  3. Compute Residual (**Double prec.** subtraction) $\Delta \mathbf{b} = \mathbf{b} - \mathbf{A}\hat{\mathbf{x}}$
  4. **Reuse** [L\U] to solve for $\Delta \hat{\mathbf{x}} = \mathbf{A}^{-1}\Delta \mathbf{b}$
  5. Compute new state estimate $\hat{\mathbf{x}}_{new} = \hat{\mathbf{x}}_{old} + \Delta \mathbf{x}$
  6. Repeat Steps 3,4,5 until

     either $\dfrac{\|\Delta \mathbf{x}\|_\infty}{\|\hat{\mathbf{x}}\|_\infty} < \varepsilon_{TOL} \ll 1$ or $\|\Delta \mathbf{x}\|_\infty \leq 10^{-d}$

Iterative Refinement fails if $K(\mathbf{A}) > 10^{+d}$ d =digits

When performing Gaussian Elimination, we have seen that the computation of an approximate condition number is almost free since we use numbers that have been computed in the normal course of the Gaussian elimination procedure. The condition number acts as a sentinel for potential RO error problems and when they indicate that the matrix **A** is ill-conditioned an *iterative refinement* procedure can be used to compute corrections to the Gaussian elimination estimate $\hat{\mathbf{x}}$ which has been compromised by the combined effect of ill-conditioning and a large number of operations.

The procedure for iterative refinement is outlined in the six steps shown on the slide; the first two just describe the usual Gaussian elimination to obtain L_1\U using pivoting and then the solution of the pair of forward and backward substitutions to find the estimate $\mathbf{x}_{hat}$. In step#3 the data residual $\Delta \mathbf{b}$ is computed and then the factorization L_1\U is reused in step#4 to solve $\mathbf{A}\,\Delta \mathbf{x} = \Delta \mathbf{b}$ for state correction $\Delta \mathbf{x}$, and step#5 updates the state estimate giving the 1st iterative refinement $\mathbf{x}^{(1)} = \mathbf{x}^{(0)} + \Delta \mathbf{x}$. Step#6 repeats steps#3-5 on $\mathbf{x}^{(1)}$ until the desired accuracy is achieved or the state correction $\|\Delta \mathbf{x}^{(k)}\|$ has a norm that is smaller than the 10^{-d} digits of precision. We note that iterative refinement fails whenever the condition number is larger than 10^{+d}.

7.8.1 Iterative Refinement Sample Calculation

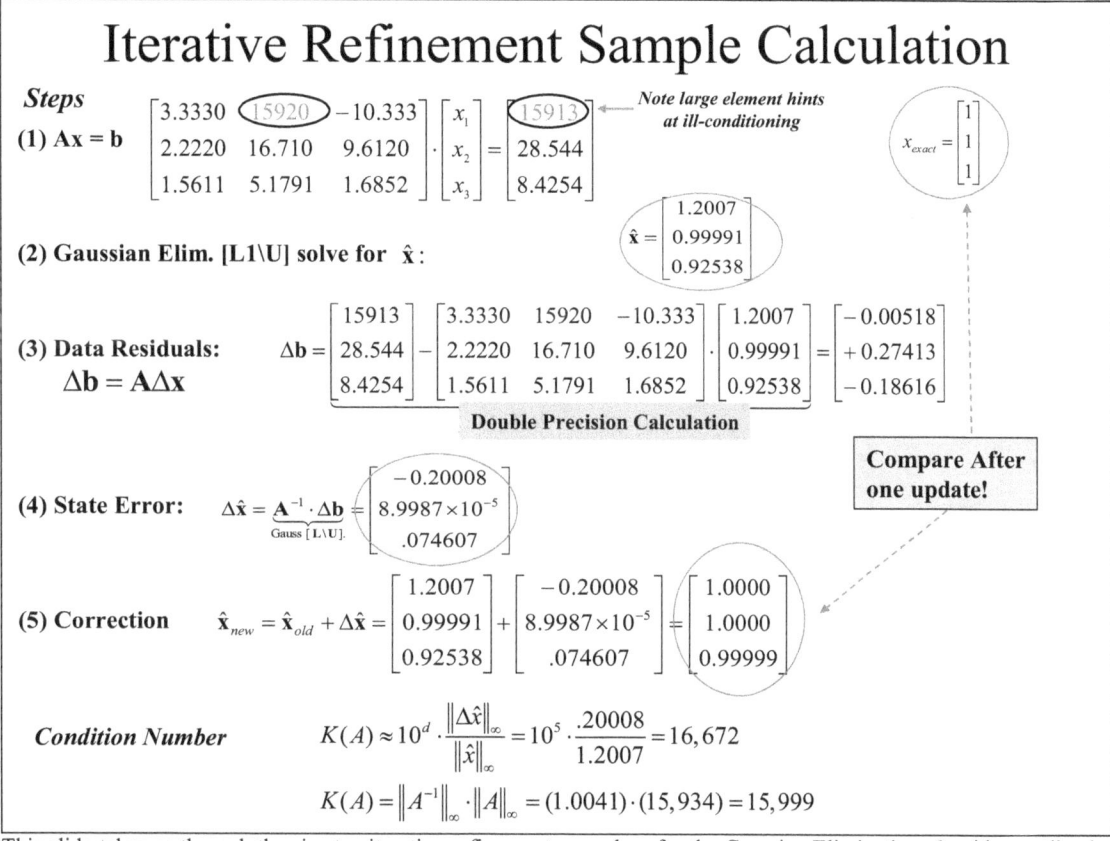

This slide takes us through the six step iterative refinement procedure for the Gaussian Elimination algorithm outlined on the previous slide. The 3x3 matrix **A** and measurement vector **b** are given on the top line with d=5 digits of precision and Gaussian Elimination yields the $L_1\backslash U$ decomposition (not shown) from which forward and backward elimination yields the Gaussian estimate x_{hat} =[1.2007, .99991, .92538]^T. The data residuals are computed using double precision and then the $L_1\backslash U$ factorization is reused to solve for the state error. The correction is made on the state to yield the 1st iterative refinement $x^{(1)}$ =[1.0000, 1.0000,.99999]^T which already compares well with the exact solution x_{exact} =[1,1,1]^T. The "Rule-of-Thumb #1" on Slide#7-34 yields a "free" condition number estimate for d=5 digits precision

$$K(A) = 10^{+d} \|\Delta x\|_\infty / \|x\|_\infty = 10^{+5} |-.20008| / 1.2007 = 16,672$$

which shows that the matrix is ill-conditioned and justifies (after the fact) the need for iterative refinement.
Had we not computed the refinement, we could still have used the other approximate condition number approximations to act as sentinels; they would tell us that the matrix is ill-conditioned and that *we should apply iterative refinement*. Also note that the K(**A**) obtained by taking the product of the infinity norms of **A** and **A**$^{-1}$ yields a value of 15,999 which is not much different than the approximation obtained from the Gaussian iterative refinement procedure.
Finally, note that the "**Rule-of-Thumb**" estimates on Slide#7-34 are validated as follows:
Rule#1): The *actual data residual ratio* is

$$\|\Delta b\|_\infty / \|b\|_\infty = (.2008/15913) = 1.7227 \times 10^{-5},$$

while *Rule#1* yields

$$\|\Delta b\|_\infty / \|b\|_\infty \approx 10^{-d} \text{ which is } \approx 10^{-5} \text{ for d= 5 digits.}$$

Rule#2): The *"true" data error ratio* is calculated from knowledge of the *exact solution* to be

$$\|\Delta x\|_{true} = \| [1,1,1]^T - [1.2001,.99991,.92538]^T \|_\infty = \max\{.2001,.00009,.07462\} = .2001$$
$$\|\Delta x\|_{true} / \|x\|_{true} = .2001/1 = .2001$$

while *Rule#2* solved for the *state error ratio* yields the upper bound

$$\|\Delta x\|_\infty / \|x\|_\infty < K_\infty(A) \cdot \|\Delta b\|_\infty / \|b\|_\infty = 15999 \cdot 1.7227 \times 10^{-5} = .27561 \text{ good upper .}$$

8 MatLab® Scripts

1. Euler Method
2. RK4 (Runge-Kutta) Method
3. RKF45 (Runge-Kutta-Fehlberg) Method
4. Adam AB4-AM3 Predictor-Corrector Method
5. Gragg Extrapolation Method
6. RK4 for Higher Order or System of ODEs
7. Jacobi and Gauss-Seidel Methods for Linear Systems
8. Successive Over Relaxation (SOR) Methods for Linear Systems
9. Comparison of Iterative Methods for Linear Systems

MatLab® Scripts

8.1 Euler Method

```matlab
%Euler
% Inputs: fty=f(t,y); Interval: [a,b], IC: y(t0)=alpha
% Outputs(6): ' index    ti      wi     y_exact    |yi-wi|   '
clear all
%********** Inputs *************
    %***** (1) ******
    %fty=inline('sin(t)+ exp(-t)', 't', 'y');
    % a=0; b=1; alpha=0; hmin=.02; hmax=.25; etol=1e-5;
    %********** y (exact) *************
    %y_exact=inline('2-cos(t)- exp(-t)', 't');
    %***** (2) ********
    %fty=inline('-t*y+4*t/y', 't', 'y');
    % a=0; b=1; alpha=1; hmin=.005; hmax=.1; etol=1e-10;
    %********** y (exact) *************
    %y_exact=inline('sqrt(4- 3*exp(-t^2))', 't');

%****** (3) *******
    fty=inline('y-t^2+1', 't', 'y');
    a=0; b=1; alpha=0.5; h=.05;%h=.2;%
    %********** y (exact) *************
    y_exact=inline('(t+1)^2-0.5*exp(t)', 't');

%********** Initialize *************
N=1+(b-a)/h
t0=a; w0=alpha; out=zeros(N,5); % Output Matrix Storage Allocation
out(1,:)=[0 t0 w0 y_exact(t0) abs(y_exact(t0)-w0) ]; % Output initial conditions "t_0" (index=1)

%********** Main *************
for i=2:N
  w0=w0+h*fty(t0,w0); % Euler step
  t0=t0+h;% Increment time by 1 h-step
  out(i,:)=[i-1 t0 w0 y_exact(t0) abs(y_exact(t0)-w0)];% Output Euler step
end
%********** Output *************
format short g
 disp('       i       ti      wi      y_exact    |yi-wi| ')
disp(out)

%********** Plots *************
plot(out(:,2),out(:,4),'b-')
hold on
plot(out(:,2),out(:,3),'rx')
plot(out(:,2),out(:,3),'r:')
plot(out(:,2),out(:,5),'g*')
plot(out(:,2),out(:,5),'g:')
grid on
title('blue Y_{exact}=(t+1)^2-0.5*exp(t);    red Euler w_i ;    green trunc error |y_i-w_i|')
%axis([0,2,0,6])
xlabel('t_i')
ylabel('y_i,  w_i,  |y_i-w_i|')
```

MatLab® Scripts

IVP: $y' = y - t^2 + 1$;
$y(0) = 0.5$; $t \in [0, 1]$; $h = .05$
$y_{exact} = (t+1)^2 - 0.5*\exp(t)$

i	ti	wi	y_exact	|yi-wi|
0	0	0.5	0.5	0
1	0.05	0.575	0.57686	0.001865
2	0.1	0.65362	0.65741	0.00379
3	0.15	0.73581	0.74158	0.005777
4	0.2	0.82147	0.8293	0.007827
5	0.25	0.91055	0.92049	0.009942
6	0.3	1.0029	1.0151	0.012123
7	0.35	1.0986	1.113	0.014371
8	0.4	1.1974	1.2141	0.016688
9	0.45	1.2993	1.3183	0.019074
10	0.5	1.4041	1.4256	0.021531
11	0.55	1.5118	1.5359	0.02406
12	0.6	1.6223	1.6489	0.026662
13	0.65	1.7354	1.7647	0.029337
14	0.7	1.851	1.8831	0.032086
15	0.75	1.9691	2.004	0.03491
16	0.8	2.0894	2.1272	0.037811
17	0.85	2.2119	2.2527	0.040787
18	0.9	2.3364	2.3802	0.043839
19	0.95	2.4627	2.5096	0.046968
20	1	2.5907	2.6409	0.050173

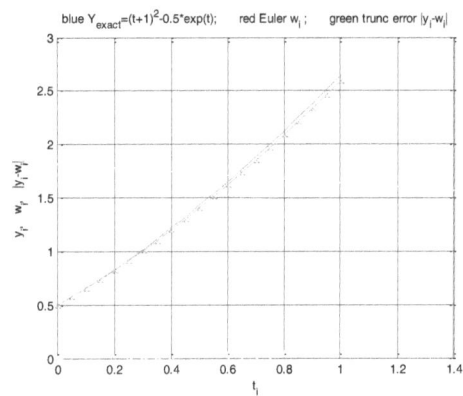

MatLab® Scripts

8.2 RK4 (Runge-Kutta) Method

```matlab
%RK4  yk=fk(t,y)
% Inputs: fty=f(t,y); Interval: [a,b], IC: y(t0)=alpha  Params: hmin, hmax etol
% Outputs(8): ' i    ti    wi    y    |yi-w1i|  '
%**********  Problem Setup  *************
% 1st order IVP  y"-3y'+ 2y = t
% y"=f(t,y,y')= t-2y +3y'
% y1=y y2=y' yields
% y2'=f(t,y1,y2)=t-2*y1+3*y2 and y1' = y2
%**********  y (exact)  *************
y_exact=inline('(t+1)^2-0.5*exp(t)', 't');
%**********  Inputs  *************
clear all
fty=inline('y-t^2+1', 't', 'y');
 a=0; b=1; alpha=0.5; h=.05;

%**********  y (exact)  *************
y_exact=inline('(t+1)^2-0.5*exp(t)', 't');

%**********  Initialize  *************
N=1+(b-a)/h
t0=a; w0=alpha; out=zeros(N,5); % Output Matrix Storage Allocation
out(1,:)=[0 t0 w0 y_exact(t0) abs(y_exact(t0)-w0) ]; % Output initial conditions "t_0" (index=1)

%**********  Main  *************
for i=2:N
k1=h*fty(t0,w0);        %RK4 Slope calculations
k2=h*fty(t0+h/2,w0+k1/2);
k3=h*fty(t0+h/2,w0+k2/2);
k4=h*fty(t0+h,w0+k3);

w0=w0+(k1+2*(k2+k3)+k4)/6;   %RK4 step
t0=t0+h;              % Increment time by 1 h-step
 out(i,:)=[i-1 t0 w0 y_exact(t0) abs(y_exact(t0)-w0)];% Output Euler step
end
%**********  Output  *************
 format short g
 disp('     i     ti     wi     y_exact     |yi-wi| ')
disp(out)

%**********  Plots  *************
plot(out(:,2),out(:,4),'b-')
hold on
plot(out(:,2),out(:,3),'rx')
plot(out(:,2),out(:,3),'r:')
plot(out(:,2),out(:,5),'g*')
plot(out(:,2),out(:,5),'g:')
grid on
title('blue Y_{exact}=(t+1)^2-0.5*exp(t);    red RK4 w_i ;    green trunc error |y_i-w_i|')
xlabel('t_i')
ylabel('y_i,  w_i,  |y_i-w_i|')
```

MatLab® Scripts

IVP: $y' = y - t^2 + 1$
$y(0)=0.5$; $t \in [0, 1]$; $h=.05$
$y_{exact} = (t+1)^2 - 0.5*\exp(t)$

Stepsize h=.25

i	ti	wi	y_exact	|yi-wi|
0	0	0.5	0.5	0
1	0.25	0.92047	0.92049	1.61E-05
2	0.5	1.4256	1.4256	3.56E-05
3	0.75	2.0039	2.004	5.90E-05
4	1	2.6408	2.6409	8.71E-05

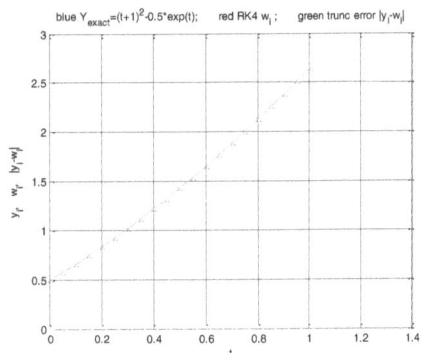

blue $Y_{exact}=(t+1)^2-0.5*\exp(t)$;   red RK4 w_i ;   green trunc error $|y_i-w_i|$

Stepsize h=.05

i	ti	wi	y_exact	|yi-wi|
0	0	0.5	0.5	0
1	0.05	0.57686	0.57686	5.20E-09
2	0.1	0.65741	0.65741	1.06E-08
3	0.15	0.74158	0.74158	1.62E-08
4	0.2	0.8293	0.8293	2.20E-08
5	0.25	0.92049	0.92049	2.80E-08
6	0.3	1.0151	1.0151	3.43E-08
7	0.35	1.113	1.113	4.08E-08
8	0.4	1.2141	1.2141	4.75E-08
9	0.45	1.3183	1.3183	5.45E-08
10	0.5	1.4256	1.4256	6.18E-08
11	0.55	1.5359	1.5359	6.93E-08
12	0.6	1.6489	1.6489	7.71E-08
13	0.65	1.7647	1.7647	8.52E-08
14	0.7	1.8831	1.8831	9.35E-08
15	0.75	2.004	2.004	1.02E-07
16	0.8	2.1272	2.1272	1.11E-07
17	0.85	2.2527	2.2527	1.20E-07
18	0.9	2.3802	2.3802	1.30E-07
19	0.95	2.5096	2.5096	1.40E-07
20	1	2.6409	2.6409	1.50E-07

MatLab® Scripts

8.3 RKF45 (Runge-Kutta-Fehlberg) Method

```
%RKF45
% Inputs: fty=f(t,y); Interval: [a,b], IC: y(t0)=alpha  Params: hmin, hmax etol
% Outputs(6): ' index    ti       wi    h    R = del_w/h  y_exact  |yi-wi|   '
%********** Inputs *************
clear all
fty=inline('y-t^2+1', 't', 'y');
   a=0; b=4; alpha=0.5; hmin=.02; hmax=.25; etol=1e-5;
%********** y (exact) *************
 y_exact=inline('(t+1)^2-0.5*exp(t)', 't');
%********** Initialize *************
t0=a; w0=alpha; h=hmax; out=zeros(11,7);
out(1,:)=[0 a w0 hmax NaN y_exact(t0) abs(y_exact(t0)-w0) ]; % Output initial conditions "step 0"
idx0=1;   flag=1;

%********** Main *************
while flag==1
k1=h*fty(t0,w0);
k2=h*fty(t0+h/4,w0+k1/4);
k3=h*fty(t0+3*h/8,w0+3*k1/32+9*k2/32);
k4=h*fty(t0+12*h/13,w0+(1932*k1-7200*k2+7296*k3)/2197);
k5=h*fty(t0+h,w0+439*k1/216-8*k2+3680*k3/513-845*k4/4104);
k6=h*fty(t0+h/2,w0-8*k1/27+2*k2-3544*k3/2565+1859*k4/4104-11*k5/40);
doh=abs(k1/360-128*k3/4275-2197*k4/75240+k5/50+2*k6/55)/h;

   if doh <= etol % Test Tolerance
   w0=w0+25*k1/216+1408*k3/2565+2197*k4/4104-k5/5;
   idx0=idx0+1;
   t0=t0+h;
   out(idx0,:)=[idx0-1 t0 w0 h doh y_exact(t0) abs(y_exact(t0)-w0)];
   end

  qstp=.84*(etol/doh)^.25; % Always compute new step ratio

   if qstp <= 0.1  % Constrain new step size
      h=.1*h;
   elseif qstp >= 4
      h=4*h;
   elseif qstp >= 0.1 && qstp <= 4
      h=qstp*h;
   end

   if h > hmax
      h=hmax;
   end

   if t0 >= b  % Finishing Step
      flag=0;  % Jump out of While flag==1 loop
   elseif t0+h > b
      h=b-t0;
   elseif h<hmin
      flag=0;
      'h less than hmin'
   end

end  %end while

%********** Output *************
 format short g
   disp('     index    ti       wi       h     R = del_w/h   y_exact    |yi-wi| ')
disp(out)
```

8-6

MatLab® Scripts

IVP: $y' = y - t^2 + 1$; $y(0) = 0.5$; $t \in [0, 4]$;
$y_{exact} = (t+1)^2 - 0.5 \cdot \exp(t)$
Variable Stepsize $h_{min} = .02$; $h_{max} = .25$; $\varepsilon_{tol} = 10^{-5}$;

index	ti	wi	h	R=del_w/h	y_exact	|yi-wi|
0	0	0.5	0.25	NaN	0.5	0
1	0.25	0.92049	0.25	6.21E-06	0.92049	1.31E-06
2	0.48655	1.3965	0.23655	4.49E-06	1.3965	2.57E-06
3	0.72933	1.9537	0.24278	4.27E-06	1.9537	4.17E-06
4	0.97933	2.5864	0.25	3.77E-06	2.5864	6.19E-06
5	1.2293	3.2605	0.25	2.44E-06	3.2605	8.50E-06
6	1.4793	3.9521	0.25	7.22E-07	3.9521	1.11E-05
7	1.7293	4.6308	0.25	1.48E-06	4.6308	1.41E-05
8	1.9793	5.2575	0.25	4.31E-06	5.2575	1.73E-05
9	2.2293	5.7818	0.25	7.94E-06	5.7818	2.07E-05
10	2.4518	6.1103	0.22244	7.99E-06	6.1103	2.45E-05
11	2.6494	6.2454	0.19762	7.12E-06	6.2453	2.87E-05
12	2.8301	6.1961	0.18068	6.63E-06	6.1961	3.33E-05
13	2.9983	5.9608	0.16821	6.35E-06	5.9607	3.85E-05
14	3.1566	5.5322	0.15828	6.17E-06	5.5322	4.43E-05
15	3.3066	4.901	0.15	6.05E-06	4.901	5.06E-05
16	3.4494	4.0563	0.14287	5.95E-06	4.0562	5.76E-05
17	3.5861	2.9863	0.13663	5.88E-06	2.9862	6.54E-05
18	3.7171	1.6784	0.13106	5.82E-06	1.6784	7.38E-05
19	3.8432	0.11973	0.12606	5.77E-06	0.11964	8.31E-05
20	3.9647	-1.7035	0.1215	5.72E-06	-1.7036	9.32E-05
21	4	-2.299	0.035323	4.82E-08	-2.2991	9.66E-05

MatLab® Scripts

8.4 Adams AB4-AM3 Predictor-Corrector Method

```
%Predictor-Corrector with RK4 generated start values
% Inputs: fty=f(t,y); Interval: [a,b], IC: y(t0)=alpha Params: hmin, hmax etol
% Outputs: ' i   ti   wi   y   h   err_test  |yi-w1i|'
%********** Inputs ************
clear all
if 1
eq='y-t^2+1';
fty=inline(eq, 't', 'y'); %need to change RK4start FUNCTION!!!
a=0; b=2.0; alpha=0.5; tol=1e-5; hmin=.01; hmax=.25;
y_exact=inline('(t+1)^2-0.5*exp(t)', 't');
end
%********** Initialize *************
h=hmax; t0=a; w0=alpha; err_testsw =1;
[w0,w1,w2,w3,t0,t1,t2,t3]= rk4start(t0,w0,h,eq);          % get RK4 start values
pc_out(1,:)=[0 t0 w0 y_exact(t0) h err_testsw abs(y_exact(t0)-w0) ]; % Output initial conditions
pc_out(2,:)=[1 t1 w1 y_exact(t1) h err_testsw abs(y_exact(t1)-w1)];%
pc_out(3,:)=[2 t2 w2 y_exact(t2) h err_testsw abs(y_exact(t2)-w2)];%
pc_out(4,:)=[3 t3 w3 y_exact(t3) h err_testsw abs(y_exact(t3)-w3)];%
j=4; contflag=1; qflag=0
%********** Main ************
while contflag==1
  wpred=w3+(h/24)*(55*fty(t3,w3)-59*fty(t2,w2)+37*fty(t1,w1)-9*fty(t0,w0));
  wcorr=w3+(h/24)*(9*fty(t3+h,wpred)+19*fty(t3,w3)-5*fty(t2,w2)+fty(t1,w1));
  err_test =19*abs(wcorr-wpred)/(270*h);
  if (err_test< .1*tol) | (err_test > tol) ;
    % change stepsize within limits
    q=(tol/(2*err_test))^.25;
    h=q*h;
       if q > 4 ;  h=4*h;  end %set stepsize ratio within limits
    if h >hmax; h=hmax; elseif h<hmin; h = hmin;disp('hmin exceeded'); end

    if j==4 %first time through loop
    [u0,u1,u2,u3,s0,s1,s2,s3]= rk4start(t0,w0,h, eq); %restart with RK4 forward
    else
    [u0,u1,u2,u3,s0,s1,s2,s3]= rk4start(t4,w3,-h, eq); %restart with RK4 backward
    end
      w0 =u0; t0= s0 ;% Overwrite prior start values
      w1 =u1; t1= s1 ;
      w2 =u2; t2= s2 ;
      w3 =u3; t3= s3 ;
      qflag=1;
       err_testsw=err_test;
  else % result is within desired tolerance
    t4=t3+h;
    w4=wcorr;
     if qflag==1
      if j==4 % first time through the while loop
        pc_out(j-2,:)=[j-3 s1 u1 y_exact(s1) h  err_testsw abs(y_exact(s1)-u1)];% w0 is fixed initial condition
        pc_out(j-1,:)=[j-2 s2 u2 y_exact(s2) h  err_testsw abs(y_exact(s2)-u2)];% Overwrite w1,w2,w3 only
        pc_out(j,:)  =[j-1 s3 u3 y_exact(s3) h  err_testsw abs(y_exact(s3)-u3)];% after a stepsize change
      end
      pc_out(j+1,:)  =[j t4 w4 y_exact(t4) h  err_testsw abs(y_exact(t4)-w4)];% Output PC integration step
      qflag=0;
    else
      pc_out(j+1,:)=[j t4 w4 y_exact(t4) h  err_testsw abs(y_exact(t4)-w4)];% Output PC integration step
    end

    j=j+1;                   % increase pred-corr output index
    w0=w1; w1=w2; w2=w3; w3=w4;       % advance all values for next step
    t0=t0+h; t1=t1+h; t2=t2+h; t3=t3+h;   % Increment times by 1 h-step
```

MatLab® Scripts

```
        if t4>b; contflag=0; end         % exit main loop
    end
end


%********** Output *************
%format short g
format short g
disp('    i      ti      wi      y_exact     h      err_test     |yi-wi| ');
disp(pc_out)
```

RK4 Starter Function

```
% This is a MatLab(C) function that must be saved as "rk4start.m" ; it is called from the main script.
function [w0,w1,w2,w3,t0,t1,t2,t3]=rk4start(t0_start,w0_start,h,eq)
    %RK4 Pred-Corr Starter
    fty=inline(eq, 't', 'y');
    %t0_start=0; w0_start=.5;h=.1;
    t0=t0_start;
    w0=w0_start;
    rk4out(1,:)=[t0,w0];         %RK4 Start values
    for j=2:4
    k1=h*fty(t0,w0);             %RK4 Slope calculations
    k2=h*fty(t0+h/2,w0+k1/2);
    k3=h*fty(t0+h/2,w0+k2/2);
    k4=h*fty(t0+h,w0+k3);
    w0=w0+(k1+2*(k2+k3)+k4)/6;   %RK4 step
    t0=t0+h;
    rk4out(j,:)=[t0,w0];
    end
    if h>0
    t0=rk4out(1,1);t1=rk4out(2,1);t2=rk4out(3,1);t3=rk4out(4,1);
    w0=rk4out(1,2);w1=rk4out(2,2);w2=rk4out(3,2);w3=rk4out(4,2);
    else
    t3=rk4out(1,1);t2=rk4out(2,1);t1=rk4out(3,1);t0=rk4out(4,1);
    w3=rk4out(1,2);w2=rk4out(2,2);w1=rk4out(3,2),w0=rk4out(4,2);
    end
```

IVP: $y' = y - t^2 + 1$; $y(0)=0.5$; $t \in [0, 1.5]$;
$y_{exact} = (t+1)^2 - 0.5 \cdot \exp(t)$

Variable Stepsize $h_{min}=.01$; $h_{max}=.25$; $\varepsilon_{tol}=10^{-5}$;

i	ti	wi	y_exact	h	err_test	|yi-wi|
0	0	0.5	0.5	0.25	0	0
1	0.125700	0.700230	0.700230	0.125700	4.051E-06	5.203E-07
2	0.251400	0.923090	0.923100	0.125700	4.051E-06	1.092E-06
3	0.377100	1.167400	1.167400	0.125700	4.051E-06	1.721E-06
4	0.502810	1.431700	1.431800	0.125700	4.051E-06	2.212E-06
6	0.628510	1.714600	1.714600	0.125700	4.610E-06	2.798E-06
7	0.754210	2.014300	2.014300	0.125700	5.210E-06	3.504E-06
8	0.879910	2.328700	2.328700	0.125700	5.913E-06	4.348E-06
9	1.005600	2.655700	2.655700	0.125700	6.706E-06	5.356E-06
9	1.108600	2.931200	2.931200	0.103010	4.819E-06	6.044E-06
10	1.211600	3.211800	3.211800	0.103010	4.819E-06	6.793E-06
11	1.314600	3.495900	3.495900	0.103010	4.819E-06	7.606E-06
12	1.417700	3.781300	3.781300	0.103010	4.819E-06	8.725E-06
14	1.520700	4.066100	4.066100	0.103010	5.326E-06	1.000E-05

8.5 Gragg Extrapolation Method

```matlab
%Gragg Method  yk=fk(t,y)
% Inputs: fty=f(t,y); Interval: [a,b], IC: y(t0)=alpha  Params: hmin, H etol
% Outputs:' i    ti    wi    y    h    |yi-w1i|  k_extrap '
%********** Inputs *************
clear all
%fty=inline('y-t^2+1', 't','y');alpha=.5;a=0;b=2;y_exact=inline('(t+1)^2-0.5*exp(t)', 't');H=.25;tol=1e-10; hmin=.02;
fty=inline('-y+t+1', 't', 'y'); alpha=1;a=0;b=1;y_exact=inline('t+exp(-t)', 't'); H=.25;tol=1e-5; hmin=.02;
%********** Initialize *************
hmax=H; t0=a; w0=alpha;
Gragg_out(1,:)=[0 t0 w0 y_exact(t0) H abs(y_exact(t0)-w0) 0]; % Output initial conditions "t_0" (index=1)
n=2;
q=[2 4 6 8 12 16 24 32];
h=H./q;
contflag=1;acceptflag=0;tableflag=1;
y=zeros(8);
for r=2:8            %compute h^2 ratio for extrapolation
   for c=2:r
Q(r,c)=(h(r-c+1)/h(r))^2 ;
   end
end
n=1; % nth row of w(t+H)table
%********** Main *************
while contflag==1
 k=1;
 z0=w0;
  while (k <= 8) && (acceptflag ==0)
    hh=h(k);nk=q(k);
    z1=w0+hh*fty(t0,w0); %Euler step
    t1=t0+hh;z0=w0;      %initialize modified midpoint values
    for j=1:nk-1
      z2=z0+2*hh*fty(t1,z1);
      if (j <nk-1) z0=z1; z1=z2; j=j+1;t1=t1+hh; end
    end
    w=.5*(z2+z1+hh*fty(t1+hh,z2));   %Averaging
    y(k,1)=w;    %first row first element of extrapolation table
     if k>=2
      r=k;
      for c=2:r %remaining elements of row "k" of extrapolation table
      y(r,c)=y(r,c-1)+((y(r,c-1)-y(r-1,c-1)))/(Q(r,c)-1);
      end
     if (k>=2) && (abs(y(r,r)-y(r-1,r-1))<=tol);  %accuracy test
        acceptflag=1;
        w0=y(r,r); %set w0 value for next iteration as last diagonal in table
        if tableflag==1   % turn on table output
          y
        end
      Gragg_out(n+1,:)=[n t0+H w0 y_exact(t0+H) H  abs(y_exact(t0+H)-w0) k]; %Gragg output
     end
    end
    k=k+1;
  end
  t0=t0+H;
  n=n+1;
  contflag=1;
  acceptflag=0;
  if(t0>b);contflag=0; end
end
%********** Output *************
 format short g
  disp('     i      ti      wi       y_exact       H       |yi-wi|       k_extrap ');
disp(Gragg_out)
```

MatLab® Scripts

IVP: f(t,y)=-y+t+1 t ; y(0) =1; t ∈ [0, 1]
Stepsize H=.25;tol=1e-5; hmin=0.02
y_{exact}= t+exp(-t);

i	ti	wi	y_exact	H	|yi-wi|	k_extrap
0	0	1	1	0.25	0	0
1	0.25	1.028801	1.028801	0.25	4.68E-09	3
2	0.5	1.106531	1.106531	0.25	7.29E-09	3
3	0.75	1.222367	1.222367	0.25	8.51E-09	3
4	1	1.367879	1.367879	0.25	8.84E-09	3
5	1.25	1.536505	1.536505	0.25	8.6E-09	3

Extrapolation Tables

1.029297	0	0
1.028927	1.028804	0
1.028857	1.028801	1.028801

1.106917	0	0
1.106629	1.106533	0
1.106574	1.106531	1.106531

1.222667	0	0
1.222443	1.222368	0
1.222401	1.222367	1.222367

1.368114	0	0
1.367939	1.367881	0
1.367906	1.36788	1.367879

1.536687	0	0
1.536551	1.536506	0
1.536525	1.536505	1.536505

MatLab® Scripts

8.6 RK4 for Higher Order or System of ODEs

```
% RK4sys %RK4sys  Higher Order ODE  yk=fk(t,y1,y2, ..., yn) and Equivalent System
% Inputs: fty=f(t,y) vector fcns; Interval: [a,b], IC: y(t0)=alpha  Params: hmin, hmax etol
% Outputs(8): ' i    ti    w1i    w2i    y    yp  |yi-w1i|  |ypi-w2i|   '
%n = order of IVP
%
%**********  Problem Setup  *************
% 2nd order IVP  y"-3y'+ 2y = t
% y"=f(t,y,y')= t-2y +3y'
% y1=y  ;  y2=y' yields
% y2'=f(t,y1,y2)=t-2*y1+3*y2 and y1' = y2

%**********   Inputs  *************
clear all
fty1=inline('y2', 't', 'y1', 'y2')
fty2=inline('t -2*y1+3*y2', 't', 'y1', 'y2')

a=0; b=3; alpha1=0;  alpha2=0.5; h=.1; n=(b-a)/h;
%**********  y (exact)  *************
yi=inline('.75*exp(2*t)-1.5*exp(t) +.5*t+.75', 't');
ypi=inline('1.5*exp(2*t)-1.5*exp(t) +.5', 't');

%**********   Initialize  *************
t0=a; w01=alpha1; w02=alpha2; %out=zeros(n+1,8);
out(1,:)=[0 a w01 w02 yi(t0) ypi(t0) abs(yi(t0)-w01) abs(ypi(t0)-w02)]; % Output initial conditions "step 0"
idx0=1;
flag=1;

%**********   Main  *************
while flag==1

k11=h*fty1(t0,w01,w02);
k12=h*fty2(t0,w01,w02);
k21=h*fty1(t0+h/2,w01+k11/2,w02+k12/2);
k22=h*fty2(t0+h/2,w01+k11/2,w02+k12/2);
k31=h*fty1(t0+h/2,w01+k21/2,w02+k22/2);
k32=h*fty2(t0+h/2,w01+k21/2,w02+k22/2);
k41=h*fty1(t0+h,w01+k31,w02+k32);
k42=h*fty2(t0+h,w01+k31,w02+k32);

w01=w01+(k11+2*(k21+k31)+k41)/6;
w02=w02+(k12+2*(k22+k32)+k42)/6;
   idx0=idx0+1;
   t0=t0+h;
   if idx0<= n+1
   y=yi(t0);
   yp=ypi(t0);
   out(idx0,:)=[idx0-1 t0 w01 w02 y yp abs(y-w01) abs(yp-w02)];
   else
   flag=0;
   end %end if

end  %end while


%**********   Output  *************
 format short g
 disp('      i      ti      w1i      w2i      y      yp    |yi-w1i|   |ypi-w2i|')
 disp(out)
```

MatLab® Scripts

```
%********** Plot ************
delw1=abs(out(:,7));
delw2=abs(out(:,8));
i1=abs(out(:,1));
plot(i1,delw1,'r-')
hold on
plot(i1,delw2,'k--')
grid on
legend(' |yi-w1i| ', ' |ypi-w2i|');
xlabel('iteration #')
ylabel('error')
hold off
```

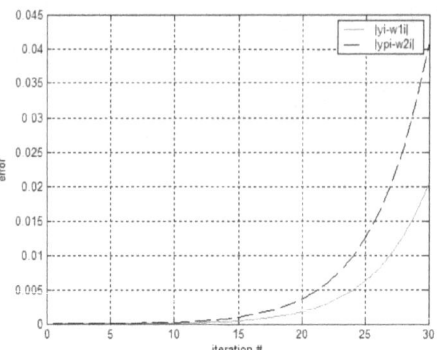

IVP: $y'' = f(t, y, y') = t - 2y + 3y'$; $y(0) = 0$; $y'(0) = 0.5$; $t \in [0, 3]$;
Equivalent System: $y_1' = y_2$ & $y_2' = y - t^2 + 1$; $y_1(0) = 0$; $y_2(0) = 0.5$;
$y_1^{exact} = .75\, e^{2t} - 1.5\, e^t + .5t + .75$; $y_2^{exact} = 1.5\, e^{2t} - 1.5\, e^t + .5$
Variable Stepsize $h_{min} = .01$; $h_{max} = .25$; $\varepsilon_{tol} = 10^{-5}$;

i	ti	w1i	w2i	yi	ypi	|yi-w1i|	|ypi-w2i|
0	0	0	0.5	0	0.5	0	0
1	0.1	0.058294	0.67434	0.058296	0.67435	1.94E-06	4.01E-06
2	0.2	0.13676	0.90562	0.13676	0.90563	4.77E-06	9.83E-06
3	0.3	0.24179	1.2084	0.2418	1.2084	8.79E-06	1.81E-05
4	0.4	0.3814	1.6005	0.38142	1.6006	1.44E-05	2.95E-05
5	0.5	0.56561	2.1043	0.56563	2.1043	2.21E-05	4.51E-05
6	0.6	0.80688	2.7469	0.80691	2.747	3.25E-05	6.62E-05
7	0.7	1.1207	3.5621	1.1208	3.5622	4.65E-05	9.45E-05
8	0.8	1.5264	4.5911	1.5265	4.5912	6.51E-05	0.000132
9	0.9	2.0477	5.8849	2.0478	5.8851	8.97E-05	0.000182
10	1	2.7142	7.5059	2.7144	7.5062	0.000122	0.000247
11	1.1	3.5623	9.5309	3.5625	9.5313	0.000164	0.000332
12	1.2	4.637	12.054	4.6372	12.055	0.000219	0.000443
13	1.3	5.9936	15.191	5.9939	15.192	0.000291	0.000587
14	1.4	7.7003	19.083	7.7007	19.084	0.000383	0.000773
15	1.5	9.8411	23.905	9.8416	23.906	0.000503	0.001013
16	1.6	12.519	29.868	12.52	29.869	0.000656	0.00132
17	1.7	15.861	37.234	15.862	37.235	0.000852	0.001715
18	1.8	20.023	46.321	20.024	46.323	0.001103	0.002219
19	1.9	25.196	57.52	25.197	57.523	0.001424	0.002862
20	2	31.613	71.31	31.615	71.314	0.001832	0.003682
21	2.1	39.563	88.276	39.565	88.28	0.002352	0.004724
22	2.2	49.398	109.13	49.401	109.14	0.003012	0.006047
23	2.3	61.548	134.76	61.552	134.77	0.003849	0.007724
24	2.4	76.543	166.22	76.548	166.23	0.004909	0.009848
25	2.5	95.03	204.83	95.036	204.85	0.006249	0.012533
26	2.6	117.8	252.2	117.81	252.21	0.007942	0.015924
27	2.7	145.83	310.27	145.84	310.29	0.010078	0.020202
28	2.8	180.29	381.45	180.3	381.47	0.012771	0.025595
29	2.9	222.65	468.66	222.66	468.69	0.016162	0.032384
30	3	274.67	575.47	274.69	575.51	0.020428	0.040925

MatLab® Scripts

8.7 Jacobi and Gauss-Seidel Methods for Linear Systems

```
%Jacobi and Gauss-Seidel Methods for Linear Systems  A*x = b
% Inputs:A, b Example: E1: 2*x1 + x2 = 2   E2:  x1 - 2*x2 = -2  ;  A=[2 1; 1 -2]; b=[2 -2]' ; Initial Guess Vector: x0 = [0 0]'
% Outputs: ' xiterate_n    '
clear all
%**********   Inputs  *************
A=[2 1; 1 -2]; b=[2 -2]' ; N=size(b,1);
%**********   Initialize  *************
x0 = [0 0]'; TOL=1e-4; testJ =1;testGS=1;
D =[2 0;0 -2]; L0=-[0 0;1 0];  U0=-[0 1;0 0];
T_J=D^(-1)*(L0+U0); C_J=D^(-1)*b;
T_GS=(D-L0)^(-1)*U0; C_GS=(D-L0)^(-1)*b;
xkGS_0 =[0 0]';xkJ_0 =[0 0]';
%**********   Main  *************
iterJ=0;  %initialize for Jacobi
while testJ>TOL
 iterJ =iterJ +1;
 xkJ=T_J*xkJ_0+C_J;
 tempJ=xkJ-xkJ_0;
 testJ = sqrt(tempJ'*tempJ);
 xkJ_0=xkJ;
end

iterGS=0;  %initialize for Gauss-Seidel
while testGS>TOL
 iterGS =iterGS +1;
 xkGS=T_GS*xkGS_0+C_GS;
 tempGS=xkGS-xkGS_0;
 testGS = sqrt(tempGS'*tempGS);
 xkGS_0=xkGS;
end

%**********   Output  *************
 format long g
'[iterJ  xkJ ]='
[iterJ , xkJ']
'[iterGS  xkGS]='
[iterGS , xkGS']

[iterJ  xkJ ]= [15         0.39996337890625         1.20001220703125]

[iterGS  xkGS]= [9         0.400009155273438         1.20000457763672]
```

MatLab® Scripts

8.8 SOR Relaxation Methods for Linear Systems

```
%Linear System  A*x = b Iterate Comparison for SOR, vary parameter omega
% Inputs:A, b Example2: E1: 4*x1 + 3*x2 = 24    E2:  3*x1 +4*x2-x3 = 30  ; E3: -x2+4*x3=-24   A=[4 3 0; 3 4 -1; 0 -1 4]; b=[24 30 -24]'
% Initial Guess Vector: x0 = [1 1 1]'
% Outputs: ' xiterate_n  '
clear all
%********** Inputs **************
A=[4 3 0; 3 4 -1; 0 -1 4]; b=[24 30 -24]' ; N=size(b,1);
%********** Initialize **************
 TOL=1e-4; testSOR=1;
imax=20;
D =[4 0 0; 0 4 0; 0 0 4]; L0=-[0 0 0; 3 0 0; 0 -1 0]; U0=-[0 3 0; 0 0 -1; 0 0 0];
x0=[1,1 1]'; xkSOR_0 =x0;
omega=[1 1.1 1.2 1.3 1.4 ]; % for output tables also for table set outflag =1
out(1,:) = [0 x0' x0' x0' x0' x0']; % for output tables also for table set outflag =1
itertbl=1;
%omega=linspace(1,2,20); %for relaxation parameter plot
 for i=1:size(omega,2)  %SOR operators for all parameters omega set up in linspace above
   T_SOR{i}=(D-omega(i)*L0)^(-1)*((1-omega(i))*D+omega(i)*U0); C_SOR{i}=omega(i)*(D-omega(i)*L0)^(-1)*b;
end
outflag =1; %No table output for table set outflat =1
%********** Main **************
for i=1:size(omega,2)
  iter=1;testSOR=1;xkSOR_0 =x0; % reinitialize for next omega value
while (testSOR>TOL)&& (iter <= imax)
 iter =iter +1;
xkSOR=T_SOR{i}*xkSOR_0+C_SOR{i};
tempSOR=xkSOR-xkSOR_0;
testSOR = sqrt(tempSOR'*tempSOR);
 if outflag==1
 out(iter,3*(i-1)+2:3*(i-1)+4) =  xkSOR';% table of iterates
  end
 xkSOR_0=xkSOR;
end
out2(i,:)=[omega(i),iter];
itertbl=max(iter,itertbl);
 end
itertbl=max(iter,itertbl);
%********** Output **************
if outflag==1
out(:,1)=[0:itertbl-1]';
format short g
 '        Iter      SOR1        SOR2         SOR3         SOR4       SOR5'
 out
 end
plot(out2(:,1),out2(:,2));hold on
plot(out2(:,1),out2(:,2),'rX'); hold off;
%title('SOR Interations \it {vs.} Relaxation Parameter \omega')
%xlabel('\omega');ylabel('#iterations  (n)');
```

Iteration# n

ω

iter	Vector Components for ω = 1.0			Vector Components for ω = 1.1			Vector Components for ω = 1.2			Vector Components for ω = 1.3			Vector Components for ω = 1.4		
0	1	1	1	1	1	1	1	1	1	1	1	1	1	1	1
1	5.25	3.8125	-5.0469	5.675	3.7431	-5.6706	6.1	3.61	-6.317	6.525	3.4131	-6.9907	6.95	3.1525	-7.6966
2	3.1406	3.8828	-5.0293	2.9444	3.8871	-4.964	2.731	3.925	-4.7591	2.5147	4.0022	-4.4021	2.3099	4.1198	-3.8794
3	3.0879	3.9268	-5.0183	3.0987	3.9398	-5.0202	3.1213	3.9781	-5.0547	3.1434	4.0538	-5.1619	3.1502	4.1865	-5.383
4	3.0549	3.9542	-5.0114	3.0398	3.9676	-5.0069	2.9954	3.9921	-4.9914	2.9045	4.0244	-4.9435	2.7441	4.0601	-4.8258
5	3.0343	3.9714	-5.0072	3.0227	3.9826	-5.0041	3.0081	3.9969	-5.0026	3.0049	4.0063	-5.0149	3.0393	3.9957	-5.0712
6	3.0215	3.9821	-5.0045	3.0121	3.9906	-5.0022	3.0012	3.9988	-4.9998	2.9924	4.0007	-4.9953	2.9888	3.9886	-4.9755
7	3.0134	3.9888	-5.0028	3.0065	3.995	-5.0012	3.0009	3.9995	-5.0002	3.0016	3.9997	-5.0015	3.0165	3.9958	-5.0112
8	3.0084	3.993	-5.0017	3.0035	3.9973	-5.0006	3.0003	3.9998	-5	2.9998	3.9998	-4.9996	2.9978	4.0001	-4.9955
9	3.0052	3.9956	-5.0011	3.0019	3.9985	-5.0003	3.0001	3.9999	-5	3.0002	3.9999	-5.0001	3.0008	4.0007	-5.0016
10	3.0033	3.9973	-5.0007	3.001	3.9992	-5.0002	3	4	-5	3	4	-5	2.9989	4.0003	-4.9993
11	3.002	3.9983	-5.0004	3.0005	3.9996	-5.0001				3	4	-5	3.0001	4	-5.0003
12	3.0013	3.9989	-5.0003	3.0003	3.9998	-5.0001							2.9999	4	-4.9999
13	3.0008	3.9993	-5.0002	3.0002	3.9999	-5							3.0001	4	-5
14	3.0005	3.9996	-5.0001	3.0001	3.9999	-5							3	4	-5
15	3.0003	3.9997	-5.0001										3	4	-5
16	3.0002	3.9998	-5												
17	3.0001	3.9999	-5												

MatLab® Scripts

8.9 Comparison of Iterative Methods for Linear Systems

```
%Linear System A*x = b Iterate Comparison for Jacobi, Gauss-Seidel, SOR, and Orthogonal Projection Methods
% Inputs:A, b Example: E1: 2*x1 + x2 = 2   E2: x1 - 2*x2 = -2  ;  A=[2 1; 1 -2]; b=[2 -2]' ; Initial Guess Vector: x0 = [0 0]'
%************************************  ************  *************  *****************************
%  **********  Alternate inputs for 3-dimensional problem  ***********  *********  ***************************
%A=[4 3 0; 3 4 -1; 0 -1 4]; b=[24 30 -24]' ; N=size(b,1);
%**********  Initialize  *************
% TOL=1e-4; testJ =1;testGS=1;testOP =1;testSOR=1;
%imax=50;
%out=zeros(imax+1,14);
%D =[4 0 0; 0 4 0; 0 0 4]; L0=-[0 0 0; 3 0 0; 0 -1 0]; U0=-[0 3 0; 0 0 -1; 0 0 0]; omega=1.25; mu=1.0
%T_J=D^(-1)*(L0+U0); C_J=D^(-1)*b;
%T_GS=(D-L0)^(-1)*U0; C_GS=(D-L0)^(-1)*b;
%T_SOR=(D-omega*L0)^(-1)*((1-omega)*D+omega*U0); C_SOR=omega*(D-omega*L0)^(-1)*b;
%*********************  ***********  **********  ************  *********  ***************  **************  ***
clear all
%**********  Inputs  *************
A=[2 1; 1 -2]; b=[2 -2]' ; N=size(b,1);
%**********  Initialize  *************
 TOL=1e-4; testJ =1;testGS=1;testOP =1;testSOR=1;
imax=20;
out=zeros(imax+1,10);
D =[2 0;0 -2]; L0=-[0 0;1 0];  U0=-[0 1;0 0]; omega=1.2
T_J=D^(-1)*(L0+U0); C_J=D^(-1)*b;
T_GS=(D-L0)^(-1)*U0; C_GS=(D-L0)^(-1)*b;
T_SOR=(D-omega*L0)^(-1)*((1-omega)*D+omega*U0); C_SOR=omega*(D-omega*L0)^(-1)*b;
ak1=[2 1]'; ak2 = [1 -2]'; P1=eye(N)-ak1*ak1'/(ak1'*ak1);P2=eye(N)-ak2*ak2'/(ak2'*ak2);T_OP={P1, P2};
ck1=ak1*b(1)/(ak1'*ak1);ck2=ak2*b(2)/(ak2'*ak2);
%*********************  ***********  **********  ************  *********  ***************  **************  ***
x0=[0,0]';
xkGS_0 =x0;xkJ_0 =x0; xkSOR_0 =x0; xkOP_0 =x0;

out(1,:) = [0 x0' x0' x0' omega x0'];
itertbl=1;
%out(:,1)=[0:imax]';
%**********  Main  *************
iter=1;
while (testJ>TOL) && (iter <= imax)
 iter =iter +1;
 xkJ=T_J*xkJ_0+C_J;
 tempJ=xkJ-xkJ_0;
 testJ = sqrt(tempJ'*tempJ);
 out(iter,2:3) = xkJ';
 xkJ_0=xkJ;
end

iter=1;
while (testGS>TOL) && (iter <= imax)
 iter =iter +1;
 xkGS=T_GS*xkGS_0+C_GS;
 tempGS=xkGS-xkGS_0;
 testGS = sqrt(tempGS'*tempGS);
 out(iter,4:5) =  xkGS';
 xkGS_0=xkGS;
end
itertbl=max(iter,itertbl);

iter=1;
while (testSOR>TOL)&& (iter <= imax)
 iter =iter +1;
 xkSOR=T_SOR*xkSOR_0+C_SOR;
 tempSOR=xkSOR-xkSOR_0;
```

MatLab® Scripts

```
    testSOR = sqrt(tempSOR'*tempSOR);
    out(iter,6:7) =  xkSOR';
   xkSOR_0=xkSOR;
 end
 itertbl=max(iter,itertbl);

 iter=1;
 while (testOP>TOL)&& (iter <= imax)
   iter =iter +1;

   xkOP=P1*xkOP_0+ck1;
   tempOP=b-A*xkOP;% data residual
   testOP = sqrt(tempOP'*tempOP);
   out(iter,9:10) =  xkOP';
   xkOP_0=xkOP;

   iter =iter +1;
   xkOP=P2*xkOP_0+ck2;
   tempOP=b-A*xkOP;% data residual
   testOP = sqrt(tempOP'*tempOP);
   out(iter,9:10) =  xkOP';
   xkOP_0=xkOP;
 end
 itertbl=max(iter,itertbl);
 %********** Output ************
 out(:,1)=[0:itertbl-1]';
 format short g
 '     Iter      Jacobi         Gauss-Seidel       SOR         omega     OrthogProj'
 Out
```

Iter	Jacobi		Gauss-Seidel		SOR		omega	Orthogonal Projection	
0	0	0	0	0	0	0	1.2	0	0
1	1	1	1	1.5	1.2	1.92	0	0.8	0.4
2	0.5	1.5	0.25	1.125	-0.192	0.7008	0	0.4	1.2
3	0.25	1.25	0.4375	1.2188	0.81792	1.5506	0	0	0
4	0.375	1.125	0.39063	1.1953	0.10606	0.95352	0	0	0
5	0.4375	1.1875	0.40234	1.2012	0.60668	1.3733	0	0	0
6	0.40625	1.2188	0.39941	1.1997	0.25468	1.0781	0	0	0
7	0.39063	1.2031	0.40015	1.2001	0.50217	1.2857	0	0	0
8	0.39844	1.1953	0.39996	1.2	0.32816	1.1398	0	0	0
9	0.40234	1.1992	0.40001	1.2	0.45051	1.2424	0	0	0
10	0.40039	1.2012	0	0	0.36449	1.1702	0	0	0
11	0.39941	1.2002	0	0	0.42497	1.2209	0	0	0
12	0.3999	1.1997	0	0	0.38244	1.1853	0	0	0
13	0.40015	1.2	0	0	0.41234	1.2104	0	0	0
14	0.40002	1.2001	0	0	0.39132	1.1927	0	0	0
15	0.39996	1.2	0	0	0.4061	1.2051	0	0	0
16	0	0	0	0	0.39571	1.1964	0	0	0
17	0	0	0	0	0.40302	1.2025	0	0	0
18	0	0	0	0	0.39788	1.1982	0	0	0
19	0	0	0	0	0.40149	1.2013	0	0	0
20	0	0	0	0	0.39895	1.1991	0	0	0

Note1: SOR method $\omega = 1.2$ is fixed; zeros in column are a table artifact

Note2: SOR columns are caught in a cycle loop with solution values oscillating in the second digit.

9 References

1. *"Scientific Computing"*, Heath, M., McGraw-Hill, Boston, 2002.
2. *"Numerical Analysis, 9h Ed."* Burden, R. L., Faires,J.D., Brooks-Cole, CA, 2011.
3. *"Introduction to Scientific Computing,"* Van Loan, C.F., Prentice-Hall, NJ, 2000.
4. *"Numerical Methods Using MatLab 4th Ed.,"* Mathews,J.H., Fink, K.D., C.F., Prentice-Hall, NJ, 2004.
5. *"Numerical Recipes in Fortran 77,2nd Ed.,"* Press, W.H.,Teukolsky, S.A., Vetterling, W.T., Flannery, B.P.,Cambridge, UK, 1996.
6. *"Numerical Analysis - A Practical Approach,2nd Ed.,"* Maron, M.J., Macmillan Publishing Co., New York, 1987.
7. *"Numerical Analysis,"* Patel, V.A., Harcourt Brace College Publishing Co., New York, 1994.

Index

A

absolute row/column sums, 7-4
accuracy tolerance ε_{Tol}, 3-9
Adams-Bashforth (AB4), 3-2
Adams-Moulton (AM3), 3-2
additive matrix decomposition, 7-21
A-h plane, 5-12, 5-13
Aitkens accelerated convergence, 7-24
algorithm efficiency, 6-3
algorithm examples, 2-24, 2-28, 3-8, 3-11, 4-6, 7–29, 7-31
alternating symbol ε_{ijk}, 6-26, 6-27
analytic solution stability, 1-7, 1-8, 1-9, 5-5
anti-symmetric product definition of determinant, 6-26, 6-27
augmented matrix [**A:b**], 6-8, 6-9, 6-10, 6-13
    with scale vector [**A:b:s**], 6-22

B

backward difference ∇, 3-3
backward Euler method, 3-9, 5-12, 5-14, 5-15, 5-16
backward substitution, 6-7, 6-9, 6-10
banded matrices, 6-17
binomial coefficients, 3-3

C

canonical coordinates, 7-7
chain rule, 2-14
characteristic polynomial, 1-11, 5-8, 5-11
Choleski $\mathbf{LL}^T$ matrix factorization, 6-29, 6-31
coefficient integrals I_k, 3-3, 3-4
cofactor, 6-26
column vectors form matrix, 6-23
complex plane, 5-9
computational efficiency, 6-16, 6-22, 7-13
    compare Euler, RK2, RK4, 2-23
    compare RK4, RKF45, 2-26
    condition number K(**A**), 7-34
    implicit methods, 5-12
    iterative methods, 6-4, 7-30
    order of method, 2-22
    pivoting, 6-21
    special matrices, 6-17
    stepsize control, 2-25
condition number K(**A**), 6-4, **7-1**, 7-36
    definition, 7-33
    estimates, 7-34, 7-35, 7-36
consistency, 2-6, 5-2, 5-10
    test, 5-3
consistency, convergence, stability, 5-2
continuity, 1-13, 5-10
convergence, 5-2, 5-10
    criteria, 7-14
    of vector sequence, 7-5
    rate, 7-23, 7-24
convex set, 1-12, 1-13
counting operations, 6-3, 6-11, 6-13
    backward and forward solution, 6-16
coupled system, 4-2, 4-3
Crout $\mathbf{LU}_1$ matrix factorization, 6-29, 6-30
Crout matrix factorization, 6-29
cubic exactness, 2-20
cumulative truncation error $\tau[h]$, 2-10, 5-4, 5-16
    down one order, 2-7
    Euler estimate, 2-8
    stepsize control, 2-25, 3-7
cutoff eigenvalue, 7-8

D

data residual vector $\Delta\mathbf{b}$, 7-1, 7-32, 7-34, 7-36
    condition number interpretation, 7-33
    iterative refinement, 7-35, 7-36
    zero out components, 7-25
data vector **b**, 6-3, 6-4, 7-32 to 7-36
determinant, 6-26, 7-9
    geometric interpretation, 6-27
    properties, 6-28
    SDD matrix, 6-33
deterministic solution of linear system, 6-2, 6-7
diagonal elements, 7-15 to 7-19
diagonal matrix **D**, 6-32, 7-21, 7-22
diagonally dominant matrix, 7-19, 7-20
difference equation (DfE), 1-11, 2-2
    consistency, 5-2
    Euler method, 2-8, 2-9
    general solution, 5-6, 5-7
    multistep methods, 3-4
    parasitic solutions, 5-6
    sampled slope method, 2-12
    vectorized form, 4-3
differential equation (DE), 2-2
    solution family plot, 1-5, 1-8, 1-9
    stiff, 5-12, 5-14
Direct Solution of Linear Systems, 6-1
domain D, 1-3, 1-12, 4-2, 5-4
Doolittle $\mathbf{L}_1\mathbf{U}$ matrix factorization, 6-29, 6-30

E

eigenvalues-eigenvectors, 7-7, 7-8
    example computation, 7-9, 7-10
    geometric interpretation, 7-7
    orthogonal eigenvectors, 7-10
    relation to 2-norm, 7-4, 7-6
    relation to spectral radius, 7-11
error bounds fo iterative methods, 7-23
error control, 2-25
Euclidean norm, 7-2, 7-6, 7-11
    distance between vectors, 7-5
Euler convergence theorem, 2-8
Euler method, 2-2 to 2-5, 2-8, 2-10, 2-11, 2-24, 8-2
    backward implicit, 3-9
    consistency test, 5-3
    modified Euler (RK2), 2-10, 2-16
    RO vs truncation error, 2-9
    stability IVP solution, 5-6
    stepsize restriction for stability, 5-11 to 5-15
    three dimensional representation, 2-4
Euler slope, 2-2, 2-3
Euler trajectory, 2-3, 2-4, 2-6
existence, uniqueness, stability, 1-2, 1-4, 1-7, 1-13, 4-2
expansion by minors, 6-26
explicit methods, 3-2, 3-5
exponential error growth, 5-7
extrapolation methods, 3-9, 3-10, 5-16

Index

F

formal solution of DE, 2-12
forward integration step, 3-2, 3-3, 3-6
forward substitution, 6-8
forward-backward operation count, 6-16
free parameter, 2-15, 2-16, 2-17
function of two variables, 2-14
functional evaluations, 2-22, 2-23, 5-16
functional iteration, 7-14

G

Gauss matrix inversion, 6-24
Gaussian elimination, 6-3, 6-7, 6-11, 6-33, 7-13, 7-30, 7-32, 7-33, 7-34, 7-35, 7-36
   algorithm derivation, 6-11
   direct matrix inverse, 6-24, 6-25
   example, 6-13
   matrix factorization, 6-14, 6-15
   operation counts, 6-12, 6-16
   rationale, 6-10
Gauss-Jordan matrix inversion, 6-25
Gauss-Seidel method, 7-16, 7-19, 7-22, 7-30, 7-31, 8-14, 8-16
   convergence theorems, 7-20
   divergence, 7-17
global truncation error, 2-7, 5-4, 5-16
Gragg extrapolation method, 3-9, 3-10, 3-11, 5-16, 8-10

H

h-A plane, 5-7
homogeneous matrix equation $(A-\lambda I)u = 0$, 7-9
HP 48G calculator plot, 1-5, 1-8, 1-9
Huen method, 2-16
hyper-volumes, 6-27, 6-28

I

identity matrix I, 7-9
ill-conditioned matrix, 6-4, 7-32 to 7-35
implicit methods, 3-2, 3-5, 5-12, 5-14, 5-15
implicit Runge-Kutta method, 5-16, 5-17
induced norm, 7-4
initial condition (IC), 2-8, 2-9, 2-20
   IVP, 1-3, 1-4
   multistep methods, 3-6
   perturbation, 1-7
   stability, 5-5
initial guess $x^{(0)}$, 6-4, 7-13 to 7-18
initial value problem (IVP), 1-3, 1-4, 1-13
   higher order methods, 2-10
   perturbed, 2-5
   solution techniques, 5-16
   standard of instability, 5-5
   standard of stability, 5-5, 5-12
   theorem on, 1-13
Initial Value Problem (IVP), 1-1
inner product, 6-8, 6-9
integer power sums, 6-12
inverse of A, 6-5, 6-6
iterative methods, 6-2, 6-4, 7-13
   accelerating convergence, 7-24
   common framework, 7-21, 7-22, 7-23
   convergence, 7-13 to 7-19
   diagonal dominance, 7-19
   divergence, 7-17
   formulation, 7-14 to 7-17
   summary, 7-30
iterative refinement, 6-4, **7-1**, 7-34, 7-35, 7-36
Iterative Solution of Linear Systems, 7-1
IVP. *See* initial value problem (IVP)

J

Jacobi method, 7-15, 7-21, 7-22, 7-31, 8-14, 8-16
Jordan elimination, 6-3

K

$K(A)$. *See* condition number $K(A)$
Kirchoff equations, 4-3

L

L_1U Doolittle factorization, 6-30
$L_1 \backslash U$ Gaussian factorization, 7-34, 7-35, 7-36
LDU matrix factorization, 6-29, 6-32
linear system overview, 6-2 to 6-6
Lipschitz condition, 1-12, 1-13, 2-3, 2-8, 4-2, 5-4, 5-10
Lipschitz constant, 2-5
LL^T Choleski factorization, 6-31
local truncation error, 2-8
   cf. cumulative truncation error, 5-4
   Euler analysis, 2-6, 2-8
   method comparison, 2-10, 5-16
   stepsize control, 2-25
   vs. cumulative truncation error, 2-7
lower triangular matrix L, 6-3, 6-8, 6-29
   determinant, 6-28
   unit diagonals L_1, 6-14
   zero diagonals L_0, 7-21, 7-22
LU_1 Crout factorization, 6-30

M

MatLab®
   Euler/RK4 comparison, 2-24
   Gragg extrapolation details, 3-11
   Jacobi, Gauss-Seidel, SOR compared, 7-31
   predictor-corrector with stepsize control, 3-8
   quiver plot, 1-6
   RK4 higher order ODE, 4-6
   RKF45 with stepsize control, 2-28
   SOR relaxation parameter study, 7-29
MatLab® Scripts, 8-1
matrix A, 7-6
   approximation technique, 7-8
   condition number $K(A)$, 7-1
   determinant calculation, 6-26
   ill-conditioned, 7-1
   is convergent, 7-12
   minor, 6-26
   pivoting, 6-18 to 6-22
   strictly diagonal dominant (SDD), 6-33
   transpose, 6-30 to 6-33
matrix decompositions, 6-3
   factorizations, 6-29, 6-30, 6-31, 6-32
   Gaussian $A = L_1 U$, 6-14, 6-15
matrix equation $Ax = b$, 7-34, 7-35
   convergence, 7-19
   formal solution, 7-21
   ill-conditioned, 7-32, 7-33
   iterative methods, 7-14 to 7-18

Index

iterative refinement, 7-35, 7-36
matrix inverse, 6-5, 6-10
   linear system solution, 6-23
   solve system of equations, 6-10
   two dimensions, 6-5, 6-6
matrix norms, 7-4
   Euclidean norm $\mathbf{A^T A}$, 7-11
   max-norm, 7-2
   p-norms, 7-2
measurement matrix $\mathbf{A}$, 6-3, 6-4
midpoint method, 2-11, 2-16
modified Euler method (RK2), 2-16
modified midpoint method, 3-9
   consistency test, 5-3
   stability IVP solution, 5-6
m^{th} order difference equation, 3-3, 5-8, 5-9, 5-13
multiple time scales, 5-12
multiplies, adds, divides (MAD), 6-11
multistep methods, 3-2, 3-4, 5-9 to 5-13
   general form DfE, 5-13
   general form of solution, 5-8
   root condition, 5-9, 5-10, 5-11
   stability, consistency, convergence, 5-6, 5-10, 5-11
Multistep Methods, 3-1

N

nesting, 2-18, 2-20, 2-21, 2-27
Newton interpolating polynomial, 3-3
Newton-Raphson method, 3-2, 3-5, 5-17
non-vetted method, 5-3
norms, 7-2, 7-3
   natural matrix norm, 7-4, 7-12, 7-33
n^{th} order Taylor method, 2-7, 5-2
numerical analysis issues, 1-2
   number of arithmetic operations, 7-13
   solution consistency, 2-6, 5-2, 5-3, 5-10
   solution existence & uniqueness, 1-10, 1-11
   solution stability, 5-5, 5-6
numerical solution method $\phi(t_i, w; h)$, 5-4

O

ordinary differential equations (ODEs), 1-2, 1-3
   equivalent system for higher order DE, 4-4, 4-5, 4-6
   vectorized form, 4-2
orthogonal projection method, 7-18, 7-22, 8-16

P

parasitic solutions, 1-11, 5-7, 5-8, 5-9, 5-13
partial derivatives, 4-2
   subscript notation, 2-19
partitioned column vector matrix, 7-8
partitioned matrix, 6-23
permutation, 6-26
perturbations, 1-11, 1-13, 2-5, 5-2, 5-5
pivoting, 6-7, 6-18
   partial pivoting (PP), 6-20, 6-21
   scale vector, 6-19, 6-20, 6-22
   SDD matrix, 6-33
p-norms, 7-2
polynomial roots, 5-8, 5-9
   characteristic equation of DfE, 5-10
   parasitic solutions, 5-7
   stability analysis, 5-6, 5-11
positive definite matrix, 6-7, 6-31, 6-32, 6-34, 7-20

predictor-corrector method, 3-5 to 3-8, 5-16, 8-8
principal axes of ellipse, 7-7

R

recursion update equation, 2-25
region of absolute stability, 5-12, 5-13
relative error, 7-33, 7-35, 7-36
relaxation parameter ω, 7-25 to 7-28
Richardson extrapolation, 3-9
RK2 method, 2-12, 2-15
   example, 2-17
   solution family, 2-16
RK4 method, 2-12, 2-24, 3-7, 8-4, 8-12
   vectorized form, 4-2
RKF45 method, 2-28, 5-16, 8-6
root condition, 5-9, 5-10, 5-11
root-finding technique, 3-2
roots greater than unity, 1-11
rotation to principal axes, 7-7
round off error (RO), 6-6
   and pivoting, 6-18
   deterministic matrix solutions, 6-3, 6-7
   increased number of operations, 5-12, 6-12, 7-13
   vs. truncation error, 2-9
row multipliers, 6-10, 6-11, 6-14
row scale factors, 7-34
row swaps, 6-18 to 6-22
rules of thumb
   condition number $K(\mathbf{A})$, 7-34, 7-35, 7-36
   down one order of h, 2-8
Runge-Kutta methods, 2-10, 2-12, 2-19, 2-20
   detailed calculation, 2-21
   formulation, 2-18
Runge-Kutta-Fehlberg method, 2-27

S

sampled slopes, 2-11, 2-12, **5-2**
   RK2 methods, 2-15, 2-16
   RK4 methods, 2-18
scale vector, 6-19
scaled pivoting, 6-20, 6-22
Schwarz inequality, 7-3
self-starting methods, 2-20, 2-25, **3-2**, 5-16
Self-Starting Single Step Methods, 2-1
sequence of iterates, 2-2, 2-6, 7-13
   convergence, 2-8, 7-19
   functional iteration, 7-14, 7-15, 7-16, 7-17, 7-18
   vector sequence, 7-2, 7-5
Simpson quadrature integral, 2-20
simultaneous displacements, 7-15
single Euler-step error, 2-8
single step methods
   stability, 5-11
   stability, consistency, convergence, 5-4
singularity of analytic solution, 1-4
slope field plot, 1-5, 1-6, 1-8, 1-9, 1-11
slope function $f(t, y(t))$, 1-3, 1-11, 2-15, 5-4
   3-dim. surface, 2-4
   number of evaluations, 2-20
solving order for **LU** factorizations, 6-30, 6-31, 6-32
SOR method, 7-24, 7-31, 8-15, 8-16
   explicit computation example, 7-27
   explicit set up, 7-26
   math details, 7-25, 7-28

Index

Special Numerical Considerations for IVP Techniques, 5-1
special types of matrices, 6-17
spectral decomposition, 7-8, 7-10
spectral radius $\rho(\mathbf{A})$, 7-6, 7-11, 7-12, 7-23
stability, 2-5, 3-6, 5-2, 5-7, 5-10, 5-12
    analysis examples, 5-11
    polynomial Q(h,A), 5-13
    SDD matrix, 6-33
    strong, weak, unstable, 5-9
stability, consistency, convergence
    multistep methods, 5-6, 5-10, 5-11
    single step methods, 5-4
start values for single/multi step methods, 2-20, 3-2, 3-6
state error vector $\Delta \mathbf{x}$, 7-1, 7-32 to 7-36
state estimate $\mathbf{x}_{hat}$, 7-32
state vector $\mathbf{x}$, 6-3, 6-4, 6-5, 7-32 to 7-36
steady state solution, 5-12, 5-14, 5-15
stepsize control, 2-25, 2-26, 3-6, 5-16
stepsize h, 2-2, 2-3, 3-3, 3-7, 5-4
    explicit algorithm h-dependence, 3-9
    exponential growth independent of h, 5-7
    max and min stepsizes, 3-10
stepsize ratio q, 2-25, 2-27, 3-7, 3-9, 3-10
stiff differential equations, 5-12, 5-14, 5-16
stopping criteria, 2-28, 3-8, 3-11, 7–29, 7-31
strictly dominant matrix, 7-20
successive displacements, 7-16 to 7-19
successive over relaxation method. *See* SOR method
sufficient condition, 1-13, 7-19
symmetric matrix, 6-31, 6-32, 7-10
system of equations, 7-13, 7-19, 8-12
    matrix linear system solutions, 7-14 to 7-18
    ODEs, 4-2, 4-3
Systems and Higher Order IVPs, 4-1

T

tableau for Gragg method, 3-9
tables and plots, 2-24, 2-28, 3-8, 3-11, 4-6, 7–29, 7-31
Taylor expansion, 2-2, 2-6, 2-12, 2-19, 5-3
    and consistency test, 5-3
    for 2 independent variables, 2-15, 2-18
Taylor method for ODEs
    and RK2 method, 2-15, 2-16
    n^{th} order truncation term, 2-13
    order n=2,3 examples, 2-14
theorems
    convergence and spectral radius $\rho(\mathbf{T})$, 7-23

convergent matrix, 7-12
Gauss-Seidel method convergence, 7-20
max-norm, 7-5
norms and spectral radius, 7-11
positive definite matrix, 6-34
SOR relaxation constant ω, 7–28
strictly diagonal dominant matrix, 6-33
time domain $t \in [a,b]$, 1-3, 1-4, 1-9
total derivative, 2-13, 2-14, 2-19
trade offs among IVP methods, 2-10, 5-16
transient solution, 5-12, 5-14, 5-15
transpose, 7-8
trapezoidal method, 2-11, 2-16, 5-12
trial solution for difference equations, 2-8, 5-6
tridiagonal matrix, 6-7, 6-17, 7-13
truncation error, 2-2, 3-3, **5-2**, 5-16
    multistep methods, 3-4
    trade against RO error, 2-9, 2-10
    zero error case, 2-20

U

uniformly spaced points, 3-3
unique solution, 4-2
unit circle, 5-9
unit triangular matrices, 6-28
unit vectors, 7-2, 7-33, 7-34
upper triangular matrix $\mathbf{U}$, 6-3, 6-7, 6-9, 6-10, 6-14, 6-29
    determinant, 6-28
    zero diagonals $\mathbf{U}_0$, 7-21, 7-22

V

vector sequence, 7-14, 7-15, 7-16, 7-17, 7-18, 7-19
vector triangle inequality, 7-3
vectorized IVP formulation, 4-4, 4-5, 4-6
vectors in numerical analysis, 1-3, 6-27, 7-2, 7-3, 7-20

W

well-posed IVP, 1-13, 5-4

Z

zero determinant geometric interpretation, 6-28

www.ingramcontent.com/pod-product-compliance
Lightning Source LLC
Chambersburg PA
CBHW081724170526
45167CB00009B/3690